水与建筑设计

国外建筑设计案例精选

水与建筑设计

（中英德文对照）

[德] 约阿希姆·菲舍尔　著
周　联　译

中国建筑工业出版社

目录 | Contents | Inhaltsverzeichnis

引言

水是人类生活中最重要的资源，无论对于区域社会和经济活动，还是对于自然和生态活动，它都是同等重要的。海水的味道、河水的奔流声以及海洋的广阔无垠，都是水与不同对象之间不同关系中令人惊奇的不同体现。但在过去，除了在游泳池和温泉浴池一类的建筑中，水极少会成为主导角色。现在不同了，设计者在工作中加入了水这一元素，产生了独特的可能性，其原因是它会引发无数的联想。之前，每样事情都像水晶一样简单明了，至少在涉及室内游泳池的设计时，这些场所就是增进健康、活动身体和体育竞赛的地方。随着时代的演变，这种不再具有吸引力的游泳池升级为带有异域植物、装饰的水上公园和主题SPA。彼得·卒姆托改变了这一切。在瑞士的瓦尔斯，他设计了一处极具质感、近乎朴素、与传统SPA和游泳池相比没有任何联系的建筑。圣地亚哥·卡拉特拉瓦则是又一个令人瞩目的例证，他在西班牙的巴伦西亚利用了计算机辅助设计程序开展工作，结果是一系列的肋形钢构使得建筑有了一种不断变化的外观。当建筑物不再有固定的水平和垂直线条，仅有的不变量就是地平线——水面。卡拉特拉瓦成功地颠覆了传统的形式法则——建筑成为动态，而水成为永久的静态。上述建筑已经成为现代建筑史中的转折点，来自当代设计的例子已经证明了光线、空气和太阳不再是制约建筑的要素。水正变得越来越流行，并逐渐演化为一种新的趋势，建筑、创作、光和水彼此之间在相互影响。例如，水在SPA和康复领域设计中成为决定性的因素，以及在滨水地带的游泳池作为生活和工作的区域。针对当前的这种变化，本书不仅作了总体性的概述，也精挑细选了一些特别项目，水在这些项目中以各种各样的方式扮演了引人注目的角色。

Introduction

Water is our most important source of life. It is equally defining for natural and ecological events, as it is for regional social and economic structures. The smell of the sea, the sound of a roaring river and the sheer infinity of the ocean are all aspects that make places on or near water so extraordinary. Yet, water rarely plays the leading role in buildings, with the exception of swimming pools and thermal baths. Here in particular, when working with this element, designers have unique possibilities, as it evokes such a variety of associations. In the past, everything was crystal clear—at least concerning the design of indoor swimming pools. They were a place to promote health, physical activity and sports. As time passed, however, this type of swimming pool ceased attracting visitors, which, in turn, promoted water parks and theme spas, complete with exotic plants and decor. Peter Zumthor changed everything. In Vals, Switzerland, he designed a highly sensual, almost austere spa, which had little to do with the traditional architectural approach to spas and swimming pools. Santiago Calatrava is another shining example. He used computer-controlled design processes to construct buildings in Valencia, Spain, whose series of skeletal steel ribs allow them to continuously change their shape. As the building has no immobile horizontal or vertical lines, the only fixed point is the horizon, the water surface. Calatrava succeeded in radically changing traditional concepts—architecture has become dynamic, while water remains permanently static. Such buildings mark a turning point in the history of modern architecture. Examples of modern plans show that light, air and sun are no longer the parameters that determine a design. Water is becoming increasingly popular, evolving into a new trend, where architecture, innovation, light and water closely interact. Examples, therefore, include water as a defining element of design in spas and wellness areas and swimming pools as work and living areas at the water's edge. This title not only gives an overview over such current developments, but also presents carefully selected projects where water plays a significant role in a number of ways.

Einleitung

Wasser ist unsere wichtigste Lebensquelle. Es ist gleichsam prägend für die natürlichen und ökologischen Gegebenheiten und die soziale und wirtschaftliche Struktur einer Region. Besonderheiten wie der Duft der Seeluft, das Rauschen des Flusses und die Unendlichkeit des Meeres verleihen Orten am Wasser ein einzigartiges Flair. Dennoch spielt Wasser in einem Gebäude selten die Hauptrolle, es sei denn, es handelt sich um ein Hallenbad oder eine Therme. Hier bietet das mit vielfältigen Assoziationen behaftete Element allerdings besondere Möglichkeiten für den Gestalter. Früher war alles klar – was die Gestaltung von Hallenbädern betrifft. Sie dienten als Ort für Sport und Bewegung und sollten die Gesundheit fördern. Irgendwann lockte das klassische Sportbad dann kaum noch Besucher, so dass Spaß- und Erlebnisbäder mit exotischen Pflanzen und anderen Südseeattributen ihren Siegeszug antraten. Doch dann kam Peter Zumthor. Er schuf in Vals ein sinnliches, fast klösterliches Thermalbad, das wenig mit herkömmlicher Bäderarchitektur gemein hat. Oder Santiago Calatrava, dessen Gebäude in Valencia dank computergesteuerten Entwurfsprozessen aus einer Abfolge von Stahlrippen konstruiert, kontinuierlich ihre Form wandeln. Es gibt keine Horizontalen und keine Vertikalen im Gebäude, und so ist der einzige Fixpunkt, der dem Betrachter bleibt, der Horizont, die Wasseroberfläche. Calatrava gelang es, das traditionelle Verständnis umzukehren: Die Architektur ist dynamisch geworden, das Wasser verharrt in ewiger Statik. So markieren diese Bauten einen Wendepunkt in der Geschichte der modernen Architektur. Beispiele moderner Planungen belegen, dass Licht, Luft und Sonne nicht mehr allein die Parameter sind, die einen Entwurf bestimmen. Wasser ist im Kommen und ein neuer Trend entsteht: Wasser als das bestimmende Gestaltungselement im Spa- und Wellnessbereich, Schwimmhäuser als Wohn- und Arbeitsraum am Wasser – allen gemeinsam ist das Zusammenspiel von Architektur, Innovation, Licht und Wasser. So bietet das vorliegende Buch einen Überblick über aktuelle Entwicklungen und eine Auswahl von Projekten, bei denen Wasser in ganz unterschiedlicher Form eine Rolle spielt.

Creations | Kreationen

创造

3deluxe 建筑师事务所

Cyberhelvetia

In the exhibiting pavilion *Cyberhelvetia* at the Expo.02 in Biel-Bienne, Switzerland, this exceptional installation of a virtual swimming pool was presented. Trade fair visitors could 'swim' in a pool without water, conceived as a place for gathering and communicating. A glowing blue glass cube, surrounded by reclining chairs, replaced the real swimming pool, so that visitors did not dive into water, but instead into different virtual realities. The varying interactions between virtual people and those truly present and digitally created creatures continuously produced new images and moods onto the projected pool surface, almost generating the impression of a living organism. A weather station on the pavilion ceiling constantly supplied data. In accordance with the weather, the pool's artificial water surface would change its appearance.

Cyberhelvetia

Im Ausstellungspavillon *Cyberhelvetia* auf der Expo.02 in Biel wurde diese außergewöhnliche Installation einer virtuellen Badeanstalt gezeigt. Als Ort der Begegnung und Kommunikation konzipiert, konnten die Ausstellungsbesucher in einem Schwimmbecken ohne Wasser „schwimmen“. Ein blau leuchtender, von Liegestühlen umgebener Glasquader ersetzte das reale Wasserbecken, in das der Besucher nicht in Wasser eintaucht, sondern in unterschiedliche virtuelle Realitäten. Gebadet wurde dabei in einer Licht- und Klangwelt. Die wechselseitige Interaktion zwischen real und virtuell anwesenden Menschen erzeugte immer neue atmosphärische Erscheinungsbilder auf der projizierten Pool-Oberfläche, so dass nahezu der Eindruck eines lebendigen Organismus entstand. Eine Wetterstation auf dem Pavillondach lieferte ständig Daten. Je nach Witterung veränderte die künstliche Wasseroberfläche des Pools ihr Aussehen.

数字瑞士

在瑞士比尔，因2002年世博会临时修建的“数字瑞士”展馆中，这一虚拟泳池装置被设想为一个供聚会和交流的场所，观众可以在这个没有水的“池子”里游泳。发光的蓝色玻璃容器代替了真正的游泳池，周围布置了躺椅，参观者不再潜入水中，而是被连接到各种不同的虚拟现实环境中。在虚拟的水池表面，虚拟的人物、真实的现在和数码生成的结果之间，多变的互动行为不断形成变化的影像和氛围，几乎出现只有有机生命才会有的感觉。一个气象设备设在展馆顶棚上，提供连续的数据。随着天气的变化，水池的人工水面呈现出不同的效果。

Filled with virtual water, the pool was soon brimming with the imaginative forms of live visitors created with the supplied technical devices and the Internet.

Das Wasserbecken war mit „virtuellem Wasser“ gefüllt. Von den Besuchern wurde es vor Ort mithilfe von technischen Geräten oder über das Internet mit fantasievollen Lebensformen bereichert.

水池以虚拟的方式体现，参观者通过技术设备和网络发出指令，池水就会以虚拟化的形式慢慢溢出。

ACQ 建筑师事务所

Dulwich Pool House

This Edwardian villa lies in a typical London suburb. The entire residential house was substantially remodeled and extended with a winter garden, a terrace and a swimming pool. Along the east side of the villa, the winter garden takes up the entire building's depth. In front of the villa's entrance, wide spans of clear glass offer privacy and protection from the wind, while also uniting the old ground floor with the new living areas. Placing the extension sideways permitted the creation of wind-protected pool grounds and a small garden area, which cannot be seen from the outside. The architecture is airy and spacious; the pool is reminiscent in its size and color of a Mediterranean construction.

Dulwich Pool House

Die viktorianische Villa befindet sich in einem typischen Londoner Vorort. Das gesamte Wohnhaus wurde aufwendig saniert und mit einem Wintergarten, einer Terrasse und einer Poolanlage erweitert. An der Ostseite der Villa erstreckt sich der Wintergarten über die gesamte Gebäudebreite. Vor dem Villenportal bilden Betonscheiben mit Glaseinschnitten eine Abschirmung und einen Windschutz, der auch das alte Erdgeschoss mit den neuen Wohnbereichen verbindet. Durch die Querstellung des Anbaus entstand ein windgeschützter Poolbereich und dahinter liegend ein zweiter, von außen nicht einsehbarer kleiner Gartenbereich. Die Architektur ist luftig und großzügig angelegt, die Poolanlage erinnert in ihrer Großzügigkeit und Farbigkeit eher an eine Anlage in Südeuropa.

达利奇水池住宅

这栋爱德华七世时代风格的别墅位于典型的伦敦市郊区。整座住宅经过了彻底改造并增加了冬季花园、门廊和一个游泳池。冬季花园在别墅的东侧，它加大了建筑的进深，在别墅的入口位置，大跨度的透明玻璃起到了保护隐私、遮挡风雨的作用，同时，它也将原有的一层大厅和新的起居生活区联成一体。扩展部分的外侧形成了泳池防风区和一个小的花园区，这个区域在外面是看不到的。建筑既通风又开阔，游泳池的尺寸和形式让人不禁联想到地中海地区的建筑。

棕榈岛建设集团

The Palm Islands

The United Arab Emirates Dubai on the Persian Gulf is known worldwide for its stunning buildings and construction projects. The *Palm* comprises three artificial island groups, which are organized in the shape of a palm. 2,000 villas, 40 luxury hotels, shopping centers and a yacht port with a total of 75 miles (120 kilometers) of sand beach are on the first Palm, Palm Jumeirah. The island is divided in three parts—the 2.5 mile (4 kilometer) long 'trunk,' a crown with 17 'palm leaves' and the 'crescent island.' The elegant residences on the islands are usually sumptuously decorated, including private pools as well as lush vegetation to protect the privacy of the individual villas.

The Palm Islands

Das Wüstenemirat Dubai am Persischen Golf ist vor allem für seine weltweit Aufsehen erregenden Bauwerke und Bauprojekte bekannt. *The Palm Islands* sind drei künstliche Inselgruppen, die in Form einer Palme angelegt sind. 2.000 Villen, 40 Luxushotels, Shoppingzentren und Jachthäfen samt 120 Kilometer Sandstrand befinden sich auf der ersten Palme, der Palm Jumeirah. Die Insel unterteilt sich in drei Abschnitte: den vier Kilometer langen „Stamm", die 17 „Palmenwedel" und den „Sichelmond". Die herrschaftlichen Anwesen auf den Inseln verfügen meist über luxuriöse Ausstattungen. Dazu gehören private Pools ebenso wie üppige Vegetation zur Abschirmung der einzelnen Villen.

棕榈岛

阿拉伯联合酋长国的迪拜城位于波斯湾，它以令人惊叹的建筑和建设项目闻名于世。棕榈岛项目包括三个人工岛屿群，它们都以棕榈树的形状进行组织设计。沿着75英里（120公里）长的沙滩建有2000栋别墅、40座奢华酒店、购物中心和一个游艇码头，第一棵棕榈树是珠美拉岛，由三个部分组成——2.5英里（4公里）长的“树干”、有17片“棕榈叶”的树冠和“新月形岛屿”。这些优美的住宅均有着奢华的装饰——包括私家泳池，而茂密的植物则用以保护这些独立别墅的私密性。

Picturesque bays with turquoise water were created between the 'palm leaves.' In order to ensure the residents are not forced to walk too much, each of the exclusive villas has an individual access to the sea.

Zwischen den Palmwedeln sind seichte Buchten mit türkisblauem Wasser entstanden. Und damit die Bewohner keine weiten Strecken zurücklegen müssen, wurde jede der exklusiven Villen mit einem Zugang zum Meer ausgestattet.

“棕榈叶片”之间是碧波荡漾的海水和奇特的海湾，为了确保居民们不必走得太远，每栋独立别墅都建有直通大海的出口。

Even though the island areas are connected by bridges on the water surface, residents and visitors can only reach the outer rings of the palms by boat. Maintaining a high degree of privacy remains a top priority.

Zwar verbinden auf dem Wasser liegende Stege die Inselbereiche untereinander, den äußeren Ring der Palme erreichen Bewohner und Besucher nur mit dem Boot. Die Einhaltung der Privatsphäre hat dabei höchste Priorität.

尽管在岛屿区之间的水面上有桥梁相连接，但当地居民和参观者乘坐船也只能到达棕榈岛的外环，开发者首要考虑的就是要保持高度的隐私性。

安藤忠雄

Langen Foundation

All of Tadao Ando's designs are characterized by his consistent use of minimalism. Simple geometrical structures and select materials, such as exposed concrete, glass and wood are the main elements of his architectural language. The *Langen Foundation* building is harmoniously embedded in the landscape, hidden behind the grassy slopes on the now defunct NATO base, a former missile-launching facility known as the *Raketenstation Hombroich,* in Neuss, Germany. Visitors walk through a cut-out in the semi-circular concrete wall, down a path bordered by a row of cherry trees and a man-made lake to the impressive ensemble. The museum is in an over 131 feet (40 meters) long concrete construction, which is enveloped by a glass cube. Two parallel concrete wings are added at a 45 degree angle. The construction, which in its unique simplicity has been able to unite Japanese elements with modern architectural morphology, is reflected in the water of a pond and seems to float, unencumbered by gravity.

Langen Foundation

Die Bauten Tadao Andos zeichnen sich durch einen konsequenten Minimalismus aus. Einfache geometrische Strukturen und ausgewählte Materialien wie Sichtbeton, Glas und Holz sind die Hauptmerkmale seiner Architektursprache. Der Bau der *Langen Foundation* ist harmonisch in die Landschaft eingebettet und liegt versteckt hinter grasbewachsenen Erdhügeln auf dem ehemaligen Gelände der NATO-Raketenstation Hombroich. Durch einen Betonbogen führt der Weg an Kirschbäumen und einem künstlich angelegten See entlang zu dem eindrucksvollen Ensemble. Das Museum ist in einem über vierzig Meter langen Betonbau untergebracht, der von einem Glaskubus umschlossen wird. Im 45 Grad Winkel dazu befinden sich zwei parallel zueinander gebaute Betonriegel. Das Gebäude, das in seiner einzigartigen Schlichtheit traditionelle japanische Elemente mit moderner Architektursprache zu verbinden vermag, spiegelt sich in dem Wasser des Teiches und vermittelt den Eindruck von Schwerelosigkeit.

朗根基金会

安藤忠雄一贯以极简主义风格塑造自己的作品，他所使用的主要建筑语言包括简单的几何构造和选择性的材料运用，像清水混凝土、玻璃和木材。位于德国诺伊斯的朗根基金会大楼以霍姆布洛伊美术馆而闻名，建筑隐藏在如今已经消失的前北约导弹发射基地的绿色草坡后，和谐地融入周边环境中。参观者通过半圆形混凝土墙的缺口，沿着一条两侧有樱桃树和人工湖的小道向下走，就会看到引人注目的建筑全貌。美术馆是一栋玻璃盒子，长度为 131 英尺（40 米）的混凝土建筑物。在主体的 45° 方向，有两条平行的混凝土结构侧翼建筑。这座建筑物以其独特的简洁方式将日本元素和现代建筑形态结合在一起，建筑倒映在水面，又如同漂浮，看起来格外的轻盈。

保罗·安德鲁建筑师事务所

National Grand Theater of China

Paul Andreu was awarded the contract for his design of a large ellipsoidal shell, which lies like a protective canopy in an artificial lake. The entry is open from both sides through a glass tunnel under the water—a cultural island surrounded by a park. It is nestled in an ornamental lake, measuring 114,829 square feet (35,000 square meters). Around eight years were needed to complete this symbol of a new China and the most important cultural institution of the country. The futurist dome is on the Chang'An avenue, near the Tiananmen Square and the Forbidden City. The complex includes an opera house, a concert hall, two theaters, as well as performance and exhibition areas. The opera house is, however, the main attraction with over 2400 seats. Constructed out of glass and titanium, it rises out of the shimmering silver see and resembles a luminescent pearl floating on water.

Chinesisches Staatstheater

Paul Andreu erhielt den Zuschlag für den Entwurf einer großen Muschel, die wie eine schützende Haube in einem künstlichen See ruht. Der Zugang erfolgt von zwei Seiten durch einen gläsernen Tunnel unter dem Wasser: eine Insel der Kultur umgeben von einem Park. Rund acht Jahre waren bis zur Fertigstellung der bedeutendsten Kultureinrichtung des Landes und Symbol des neuen Chinas nötig. Der futuristische Dom liegt an der Avenue Chang'An und damit nur rund 500 Meter vom Tiananmen-Platz und der Verbotenen Stadt entfernt. Der Komplex beherbergt eine Oper, eine Konzerthalle, zwei Theater sowie Veranstaltungssäle und Ausstellungsräume. Highlight ist jedoch die Oper mit über 2.400 Sitzplätzen: aus Glas und Titan erhebt sie sich aus dem silbern glänzenden See und wirkt wie eine Perle auf dem Wasser.

中国国家大剧院

保罗·安德鲁因其设计的一座位于人工湖中、有着巨大椭圆形保护盖的贝壳式建筑而受人关注。这是一个周边环绕着公园的文化之岛——建筑通过水下的玻璃隧道在两端均设有出入口，它被妥当地安放在一个面积约114829平方英尺（35000平方米）的优美湖泊中。建好这个中国新的标志和国家最重要的文化设施大约需要8年时间。这一未来主义风格的圆形顶盖位于长安街，靠近天安门广场和紫禁城。建筑综合体包括一座歌剧院、一座音乐厅和2个剧场，既可用于表演也可用于展示最重要的歌剧院拥有超过2400个座位。大剧院的外表以玻璃和钛板构成，它反射着银色的光泽，如同一颗闪闪发光的珍珠浮在水上。

The glass shell allows glimpses into the interior and a stunning view of the 114,829 square feet-sized (35,000 square meters) lake.

Die Glashülle gibt den Blick von Innen nach Außen auf den 35.000 Quadratmeter großen See frei.

玻璃外壳的设计使得室内空间和令人震撼的114829平方英尺（35000平方米）水景相互贯通。

The titanium skin protects and conceals the interior, creating secretive, shadowy areas.

Die Titanhaut beschützt und verbirgt das Innere und kreiert schattenhaft, geheimnisvolle Zonen.

钛合金表皮保护并遮蔽了室内空间，令建筑内部变得神秘和幽深。

戴水道景观设计公司

Heiner-Metzger-Platz

Atelier Dreiseitl has gained a national and international reputation for water and fountain design projects. In Neu-Ulm, Germany, the demolition of the former train station enabled the renovation of the *Heiner-Metzger-Platz*. In the transformation to an inner urban park with meeting and play points, rainwater was to play a larger role as well. The focal point of the design is a walk-through fountain that is directly experienced. Together with the pavement, the water creates one level, and the busy water surface offers lively optical stimulus. Metal constructions, which rise out of the basin, transport delicate water mists. The entire rainwater is collected, filtered and led under the pavement slabs. The fountain and water curtains only use groundwater.

Heiner-Metzger-Platz

Das Atelier Dreiseitl hat sich national wie international vor allem mit der Gestaltung von Wasser- und Brunnenprojekten einen Namen gemacht. In Neu-Ulm ergab sich durch den Abriss des ehemaligen Bahnhofs die Möglichkeit zur Neugestaltung des Heiner-Metzger-Platzes. Bei der Verwandlung zu einem innerstädtischen Park mit Spiel- und Begegnungspunkten sollte auch dem Regenwasser eine größere Bedeutung zukommen. Zentraler Blickfang der Platzgestaltung ist ein erlebbarer und begehbarer Brunnen. Wasser bildet zusammen mit dem Pflaster ein Niveau und erzeugt durch die unruhige Wasseroberfläche lebendige optische Reize. Metallkonstruktionen, die sich aus dem Becken erheben, tragen filigrane Wasserschleier. Das gesamte Regenwasser wird gesammelt, gefiltert und unter die Gehwegplatten geleitet. Der Brunnen und die Wasservorhänge werden ausschließlich aus Grundwasser gespeist.

海纳尔–梅茨格广场

戴水道公司以水景和喷泉设计享誉全球。在德国新乌尔姆，随着火车站的拆除，重建海纳－梅茨格广场成为可能。在将其改造为聚会、休闲、城市游戏广场的过程中，雨水扮演了一个重要的角色。设计的重点是一条喷泉水道的处理。水和铺装一同构成了相同的标高区域，热闹的水面形成生动的视觉亮点。金属结构支撑着蓄水池溢出美妙的水幕。所有的雨水经收集、过滤后引入铺装层下部，喷泉和水幕只用这里的地下水。

BUCHHAN

Queens Botanical Gardens

The botanical gardens in the borough of Queens in New York City contain a unique universe of plants and plant varieties. Visitors experience the diversity of flowers, bushes and trees. Exhibits, educational and research programs advocate a more responsible method of using light, air, soil and water. The extension was built with the newest environmentally friendly technologies. The roof was conceived as a collection pool for rainwater. From here, it is led over concrete grooves into different parts of the Queens Botanical Gardens. The water flows over wood, water stairs and boulders, ending in a pond, which is architecturally reduced. A small bridge leads over the water surface and unites the main terrace with the new house for Horticulture / Maintenance Building.

Queens Botanical Garden

Der Botanische Garten im Stadtteil Queens in New York beherbergt eine einzigartige Pflanzenwelt. Hier erlebt der Besucher die Artenvielfalt von Blumen, Sträuchern und Bäumen aus aller Welt. Ausstellungen, Bildungs- und Forschungsprogramme präsentieren einen verantwortlichen Umgang mit Licht, Luft, Erde und Wasser. Der Anbau wurde mit den neuesten umweltfreundlichen Technologien ausgestattet. Das als Wasserschale konzipierte Dach dient als Auffangbecken für das Regenwasser. Von hier aus wird es über Betonrinnen in verschiedene Teile des Botanischen Gartens geleitet. Der mit Holz, Wasserstufen und Findlingen eingefasste Wasserlauf endet in einem Teich, der sich architektonisch reduziert präsentiert. Ein schmaler Steg führt über die Wasserfläche und verbindet die Hauptterrasse mit dem neuen Haus für „Gartenbau und nachhaltiges Wirtschaften“.

皇后区植物园

纽约皇后区的植物园是一个独特的植物世界，参观者在这可以参观并体验花卉、灌木和树木的多样性。针对光、空气、土壤和水资源的使用展览、培训和课题研究倡导更负责的方式。扩展部分的建造使用了最新的环保技术，建筑的屋顶部分被作为雨水收集池。雨水从这里通过混凝土水道流向植物园的不同区域，水流过树林、水梯田、卵石滩，最后汇聚在一个建筑化的水池，水面一座悬空的小桥把主楼和用于园艺和维护的新建筑联系在一起。

Subtle rooms and the light axis increase the understated effect of the water course.

Subtile Räume und Lichtachsen verstärken die reduzierte Wirkung der Wasserläufe.

精心布置的建筑、某种程度的轴线关系加强了水道的平淡效果。

There are places where the elements sun, wind and, of course, water are highlighted.

Es entstanden Orte, an denen die Elemente Sonne, Wind und natürlich Wasser im Vordergrund stehen.

还有一些太阳和风的元素，当然，水是焦点。

贝尼施建筑师事务所

Spa Baths

The white domes of the spa in Bad Aibling soar over the surrounding landscape like over-dimensional water bubbles. The small German resort, some 37 miles (60 kilometers) southeast of Munich, boasts the oldest 'moor' spas in Bavaria. With their daring, futurist design, Behnisch architects complemented the old facilities with a modern spa, the so-called *Kabinettbad*. A series of different-sized theme domes, which have a diameter of 13 to 52 feet (4 to 16 meters), are grouped around an organically formed pool, which extends to a spacious winter garden and outer terraces. The different 'cabinets' resemble caves thanks to their dome-shaped roofs. Semi-round entrances and windows allow arresting views of the surrounding water.

Therme

Wie überdimensionale Wasserblasen ragen die weißen Kuppeln der Badelandschaft Bad Aibling aus der umgebenden Landschaft. Der kleine Ort, 60 Kilometer südöstlich von München ist das älteste Moorheilbad Bayerns. Der kühne, futuristische Entwurf von Behnisch Architekten ergänzt die alten Anlagen um ein modernes Thermalbad, das so genannte Kabinettbad. Es handelt sich dabei um Themenkuppeln, die einen Durchmesser von 4 bis 16 Meter haben und sich um einen organisch geformten Pool gruppieren, der sich zu einem großzügigen Wintergarten und Außenterrassen öffnet. Die verschiedenen Kabinen wirken durch ihre kuppelförmigen Dächer wie Höhlen. Halbrunde Eingänge und Fenster lassen immer wieder reizvolle Einblicke in die Wasserlandschaft zu.

Spa疗养池

白色的圆形屋顶耸立在巴特艾布灵的景色中，它们看起来像一个个超大尺度的大水泡。这个小旅游点距慕尼黑大约 37 英里（60 公里），拥有巴伐利亚最古老的“露天”矿泉疗养池。贝尼施建筑师事务所以其大胆的、带有未来主义风格的设计完成了原有设施的改造，我们称这个现代 Spa 为“Kabinettbad”。直径从 13 到 52 英尺（4~16 米）的一系列绝妙的圆形屋盖簇拥在一个形态自由的矿泉池周围，并延伸出一个开阔的冬景花园和一处室外平台。圆形顶盖让这些不同的小房间如同“洞穴”，半圆形的入口和窗洞保证了足够的视野。

恩斯特·贝内德

Haus H.

This single-family house on the Aussee Lake in Blindenmarkt, Austria, is an extended cube with a stringer shell, whose shorter side faces an artificial lake. The glass front rises over the water, offering an amazing view. The interior of the house consists of one Asian-like room, facing the water. The living area is slightly raised by a few stairs; a stairwell leads to the upper floor and the sleep gallery. A clearly defined courtyard with pebbles leads to the ground floor of the house. Large sliding glass elements allow access thereto; the side facing the courtyard makes ample use of glass façades. A separate stairwell leads to the roof terrace, which is hidden from view.

Haus H.

Das Wohnhaus am Ausee in Blindenmarkt ist ein in die Länge gezogener Quader mit Längsholzverschalung, der mit seiner Schmalseite zu einem künstlich angelegten See liegt. Die verglaste Stirnseite ragt über das Wasser und bietet einen traumhaften Blick. Das Innere des Hauses besteht aus einem einzigen, asiatisch eingerichteten Raum, der zum Wasser hin orientiert ist. Der Wohnbereich ist durch ein paar Stufen angehoben, eine Treppe führt in das Obergeschoss mit Schlafgalerie. Ein Hof mit Kieselsteinen führt zum Erdgeschoss, das zur Hofseite großzügig verglast und durch große Glasschiebeelemente zu öffnen ist. Über eine separate Treppe gelangt man auf die Dachterrasse, die vor Einblicken geschützt ist.

H住宅

这个独户住宅建在奥地利布林登马克特的奥苏湖，这是一个有着奇怪外表的出挑立方体，建筑的短边朝向一个人工湖泊。前方的玻璃窗高悬在水面上，视线非常开阔。正对水面的内部空间装饰成亚洲风格，起居区域通过几个台阶被局部抬高，楼梯通往上层和睡房。铺有卵石的简洁小院通向一层的房屋，大面积的滑动玻璃门提示了入口的位置，朝向院子的一边丰富了建筑立面。有独立的楼体通向屋顶平台，那里具有相当的私密性。

The building attracts considerable attention because the architecture in this area tends to be more influenced by the alpine landscape. With its cubic form, it resembles a cargo container more than it does a vacation house. This impression is reinforced by the reduced use of materials.

In dem architektonisch eher alpenländisch geprägten Gebiet sticht das Gebäude augenfällig heraus. Mit seiner kubischen Form erinnert es vielmehr an einen Überseecontainer als an ein Ferienhaus. Unterstrichen wird dieser Eindruck noch durch den reduzierten Einsatz der Materialien.

因为这一地区的建筑更多是受阿尔卑斯地域风格的影响，它的建成吸引了相当多的关注。它立方体般的外观，看起来不像度假别墅，反倒更像是个集装箱货柜。造成这一结果的原因是设计者使用的材料比较单一。

马里奥·博塔

San Carlino

Mario Botta is one of the leading architects from Tessin, Germany. His buildings, from his early one-family houses to the monumental constructions of the present, are generally based on few geometrical figures, such as cubes or cylinders with striking spatial incisions. No contemporary architect has built more churches than Mario Botta. An exhibition on the baroque architect Francesco Borromini scheduled in honor of his 400th anniversary was the incentive for Mario Botta to build a 108 feet (33 meters) high wooden model of the famous Roman church *San Carlo alle Quattro Fontane* along the shore of Lake Lugano. Like a mirror, the sea surface complements the deliberately omitted building shell, while the fascinating interaction of light and shadow highlights the extraordinary design. Water as a symbol of life—this is how the model designed together with Botta's architecture students was to be interpreted.

San Carlino

Mario Botta gilt als Star unter den Tessiner Architekten. Seine Bauwerke, beginnend von den Einfamilienhäusern seiner Frühzeit bis zu den Monumentalbauwerken der Gegenwart basieren grundsätzlich auf wenigen geometrischen Figuren wie Würfel oder Zylinder mit markanten, räumlichen Einschnitten. Kein zeitgenössischer Architekt hat mehr Gotteshäuser gebaut als Mario Botta. Aus Anlass der Ausstellung zum 400. Geburtstag des Barockarchitekten Francesco Borromini errichtete Botta am Ufer des Luganer Sees ein 33 Meter hohes Holzmodell der berühmten Kirche *San Carlo alle Quattro Fontane* in Rom. Wie ein Spiegel ergänzt die Seeoberfläche die bewusst weggelassene Gebäudehülle und das faszinierende Spiel von Licht und Schatten setzt eindrucksvolle Akzente. Wasser als Symbol des Lebens – so soll das, zusammen mit seinen Architekturstudenten entstandene Modell zu verstehen sein.

圣卡利诺

马里奥·博塔是建筑师中的领军人物之一，他来自德国泰辛。从早期的家庭住宅到具有纪念意义的当代建筑，其建筑风格总体上基于一些几何图形的演化，例如，有着明显空间缺口的圆柱或立方体。在当代建筑师中，他设计的教堂数量最多。在纪念巴洛克建筑师弗朗切斯科·普罗密尼诞辰 400 周年活动的展会上，马里奥·博塔受邀设计一件作品，他在卢加诺湖畔以著名的罗马教堂圣卡洛·阿勒·夸特罗·丰坦为原型完成了一座高 108 英尺（33 米）的木构模型。水面像镜子一样将设计师有意省略掉的那部分建筑体补全，迷人的光影产生的互动诉说了这一非同凡响的设计，这是博塔和他的学生们创作这一作品的注解——水是生命的象征。

Spa Bergoase

In Arose, one of the most popular and known ski and relaxation areas of Switzerland, Mario Botta created an unrivalled work of architecture—the wellness center *Bergoase,* Mountain oasis, of the five-star *Tschuggen Grand Hotel.* One of the impressive features is the 42 feet (13 meters) high light sails. Only select materials were used for the wellness area—untreated and smoothly polished Duke-White-Granite from Domodossola, glass and Canadian maple. The upper of the four areas is dedicated to the subject of water and is bathed in invigorating light, thanks to the light and shadow games through the skylights. A wave-shaped wall out of untreated granite stones limits the four inner pools, which optically merge to form one giant pool. The light stone lets the water shimmer light blue. The pools are connected with the shower areas located in the so-called Arosa-grotta, where visitors can experience the seasons.

Spa Bergoase

In Arosa, einem der beliebtesten Ski- und Erholungsgebiete der Schweiz, schuf Mario Botta eine Architektur, die ihresgleichen sucht: das Wellnesszentrum *Bergoase* des Fünf-Sterne *Grand Hotel Tschuggen.* Markanter Blickfang sind die bis zu 13 Meter hohen Lichtsegel. Für den Wellnessbereich verwendetet der Architekt nur ausgesuchte Materialien: roh belassenen sowie glatt geschliffenen Duke-White-Granit aus Domodossola, Glas und kanadischen Ahorn. Die oberste der vier Wellnessbereiche ist ganz dem Thema Wasser gewidmet und wird durch die Licht- und Schattenspiele der Oberlichter in spannungsvolles Licht getaucht. Eine wellenförmige Wand aus rohen Granitsteinen begrenzt die vier Innenbecken, die optisch zu einem großen Wasserbecken zusammenlaufen. Der helle Stein lässt das Wasser türkisblau schimmern. Eine Erlebnisdusche in der so genannten Arosa-Grotte, der den Besucher die Jahreszeiten durchleben lässt schließt daran an.

山中绿洲Spa

瑞士阿罗萨是最为流行和知名的滑雪、休养圣地，马里奥·博塔在此创造了一座无与伦比的建筑——Bergoase康复中心，意为“山中绿洲”，它是五星级酒店Tschuggen Grand Hotel的一部分。高42英尺（13米）的“光帆”是引人注目的特征之一。建筑师选择的建筑材料包括：产自多莫多索拉的未经处理和打磨光滑的杜克白花岗石、玻璃和加拿大枫木。光从天窗撒下，上部四个专门以水为主题的区域充满生气。波浪形的毛石花岗岩墙体围合而成的四个内部水池，在视觉上形成一个巨大的区域。石材的反光使得池水闪烁着微微的蓝色，水池与所谓的阿罗萨洞穴淋浴区相连，客人能直接感受到外部的四季变化。

斯蒂芬·布劳恩费尔斯建筑师事务所

Paul-Löbe-Haus

Together with the *Kanzleramt* and the Marie-Elisabeth-Lüders-Haus, the Paul-Löbe-Haus belongs to the Federal Band, which connects both parts of once-divided Berlin, crossing across the Spree River. The airy new construction designed by architect Stephan Braunfels is located directly at the arch, where the river is crossed, known as the *Spreebogen.* And right where the building borders onto the Spree shore, a waterside promenade seems to be a direct invitation for a stroll along the waterfront. The building's transparency begins at the main entrance to the west. A huge glazed surface reflects the Parliament Building across the street, the *Reichstagsgebäude,* and the movements of the Spree River. In the evening, when the glass surface is lit up from inside and the right and left symmetrically winding inner stairs show their sculptural effect, visitors experience the truly magical charm of the building.

Paul-Löbe-Haus

Das Paul-Löbe-Haus gehört neben dem Kanzleramt und dem Marie-Elisabeth-Lüders-Haus zum Ensemble „Band des Bundes", das die beiden früher durch die Mauer getrennten Teile der Hauptstadt über die Spree hinweg verbindet. Der lichte Neubau des Architekten Stephan Braunfels liegt direkt am Spreebogen. Und dort, wo das Gebäude an die Spree grenzt, lädt eine Uferpromenade zum Spaziergang ein. Die Transparenz des Hauses beginnt bereits am westlich gelegenen Haupteingang. Eine riesige verglaste Fläche spiegelt das gegenüberliegende Reichstagsgebäude und die Spree wider. Am Abend, wenn die Glasfläche von innen beleuchtet ist, und die rechts und links symmetrisch aufsteigenden Innentreppen ihre skulpturale Wirkung entfalten, wird der Besucher geradezu magisch angezogen.

保罗–吕博大楼

保罗–吕博大楼、总理府以及玛丽–伊丽莎白–吕德斯大楼都属于联邦带形区，它横跨施普雷河，将一分为二的柏林连在一起。这座轻盈的建筑由斯蒂芬–布劳恩费设计，它恰好位于河流的转折部位——所谓的施普雷湾。建筑边缘部分紧邻施普雷河，在滨水步道散步令人十分惬意。西向的主入口展示着建筑的通透特征，巨大的镀釉玻璃映衬着对面的国会大厦以及缓缓流过的施普雷河。入夜时分，当玻璃被灯光照亮，内部左右对称的盘旋楼体呈现出雕塑般的效果，游客将体会到建筑的迷人魅力。

BRT 建筑师事务所

Dockland

At the head of the Edgar-Engelhard wharf, between the northern part of the Elbe River and the fishing port, one of Hamburg's most unusual office buildings can be found. In order to build the steel and glass construction, a headland was created in the harbor basin. Like a gigantic luxury cruise ship, the building now lies on the Elbe River. The bow of the ship-like construction juts 132 feet (40 meters) out of the Elbe River and seems to float on the water. The middle area of the spacious complex contains ample place for communication zones such as small kitchenettes or conference rooms, as well as for archives or printing rooms. In the surrounding areas, office and work rooms were created; their decor and furniture can be easily changed and adapted. From the inside and thanks to the large-scale glazed façades, the view out onto the Elbe River is simply stunning, as if Dockland were truly directly on the water.

Dockland

Am Ende des Edgar-Engelhard-Kais zwischen Nordelbe und Fischereihafen steht das wohl ungewöhnlichste Bürohaus der Hansestadt Hamburg. Für die Stahl- und Glaskonstruktion wurde extra eine Landzunge im Hafenbecken aufgeschüttet. Wie ein gigantischer Luxusliner liegt das Gebäude auf der Elbe. Der „Bug" der schiffsartigen Konstruktion schwebt 40 Meter frei über der Elbe. Der Mittelbereich des großzügig bemessenen Komplexes bietet ausreichend Platz für Kommunikationszonen wie Teeküchen oder Besprechungsräume, aber auch für Archive oder Druckerräume. Die verbleibenden Freiflächen wurden mit individuellen Büro- und Arbeitsräumen ausgestattet. Im Inneren gewährt die großflächige Verglasung eine atemberaubende Aussicht auf die Elbe und man gewinnt den Eindruck, mitten auf dem Wasser zu sein.

杜克角大厦

在埃德加 – 恩格尔 – 哈德港湾的前端，易北河北部和渔港之间，你将发现一栋非同寻常的汉堡市办公建筑。为了这座钢铁和玻璃结构大楼，港口填海造了一个岬角。建筑就像一艘易北河上的超级豪华游轮，“船”头高 132 英尺（40 米），高高探出在水上。建筑巨大而复杂，中心区域有众多空间用于活动和交流，像餐饮区、会议室、档案室以及打印室等。在建筑的周边设计了办公室和工作用房，房间的装饰和家具可以很方便地进行调整和改变以适应不同需要。从建筑内部透过巨大的玻璃窗向外眺望，易北河令人惊叹的壮观景色尽在眼前，让人感觉到杜克角大厦如同行驶在河面之上。

The spacious stairwell, which is open to the public leads to the roof terrace, famous for its spectacular views.

Die großzügige, öffentliche Freitreppe führt zur Dachterrasse, die einen spektakulären Ausblick ermöglicht.

对公众开放的巨大的楼梯通往屋顶的平台，那里以其独特的港区景观闻名。

Cool colors and materials, such as steel, concrete and glass create a stark contrast to the surrounding historical buildings and give the office building its futuristic appearance.

Kühle Farben und Materialien, wie Stahl, Beton und Glas, bilden einen starken Kontrast zu den umliegenden historischen Bauwerken und geben dem Bürohaus sein futuristisches Aussehen.

偏冷的色彩，冷色系的钢材、混凝土和玻璃等材质的应用使得建筑和周边环境产生了强烈的反差，这座办公楼体现着一种未来主义风格。

圣地亚哥·卡拉特拉瓦

City of Arts and Sciences

La Ciudad de las Artes y las Ciencias, the City of Arts and Sciences, situated in the former riverbed of the Turia River in Valencia is a unique building complex comprising an opera and music palace, a 3D-cinema, a parking garage, a science museum and the largest aquarium in Europe. Inspired by human and animal forms and completely in white, this building has become a shining trademark of this Spanish city. Depending on the location, the buildings take on different appearances–sometimes as an antique helmet with an open visor, sometimes as Darth Vader, dressed in white or even as a whale's torso. In memory of the original course of water, huge water basins were constructed, in which the buildings appear as islands. Broken white tiles cover far more than just the façades, they can also be found on the floors of the basin, the flowing corners and the organically shaped benches.

Stadt der Künste und der Wissenschaften

La Ciudad de las Artes y las Ciencias, die Stadt der Künste und der Wissenschaften im ehemaligen Flussbett des Rio Turia in Valencia, ist ein einzigartiger Gebäudekomplex bestehend aus einer Oper und Musikpalast, einem 3D-Kino, einer Parkanlage, einem Wissenschaftsmuseum und dem größten Aquarium Europas. Inspiriert durch menschliche oder tierische Formen und ganz in weiß gehalten, ist das Ensemble zum strahlenden Wahrzeichen der spanischen Stadt geworden. Je nach Standort erscheinen die Gebäude mal als antiker Helm mit aufgeklapptem Visier, mal als Darth Vader in Weiß oder mal als Rumpf eines Walfischs. Als eine Erinnerung an den ursprünglichen Wasserverlauf wurden riesige Wasserbecken angelegt, in denen die Gebäude wie Inseln liegen. Weiße gebrochene Fliesen bedecken die Fassaden, die Böden der Becken, die Umrandungen und die Sitzbänke.

艺术与科学城

La Ciudad de las Artes y Las Ciencias——艺术与科学城，坐落在巴伦西亚图里亚河的河滩地上，它是独一无二的综合建筑群，包括有一个歌剧院、一个音乐厅、一个 3D 影院、停车场、一个科学博物馆和欧洲最大的水族馆。设计的灵感来自于人和生物的形态，全白的建筑群成为这座西班牙城市引人注目的地标。从不同的角度看，它有着不同的外观——有时像古代打开了帽舌的头盔，有时像全身着白的黑武士达思・韦德（星球大战中的反派主角），甚至像巨鲸的骨架。设计者为纪念原有河道而设计了巨大的水池，建筑如同岛屿一样坐落其中。白色的瓷砖碎片除了用在建筑立面外，还用在水池的底层、池壁和长凳上。

The light that streams through the ribs of the science museum, *Museo de las Ciencias Príncipe Felipe,* is reflected in the water where it throws mysterious shadows.

Das Licht, das durch die Rippen des Wissenschaftsmuseums, *Museo de las Ciencias Príncipe Felipe,* dringt, spiegelt sich im Wasser und zaubert geheimnisvolle Schatten.

光透过科学馆的肋形骨架，在水中投下变幻的倒影。

The 3D-movie theater *L'Hemisfèric,* which is conceived as an opening-and-closing eye, seems to float over the water surface.

Das 3D-Kino *L'Hemisfèric,* das als sich öffnendes und schließendes Auge konzipiert ist, scheint über der Wasseroberfläche zu schweben.

3D电影院像不断开阖的眼睛漂浮在水上。

蓝天组设计事务所

Art Museum

The Art Museum in the city of Groningen, in the Netherlands, is not only remarkable for its art. Even the building complex is well worth a visit. The concept for the pavilions was made by the Italian designer Alessandro Mendini in collaboration with Michele de Lucchi, Philippe Starck and Coop Himmelb(l)au. The museum lies on an island in the middle of the water in the connecting channel of the city. Two pavilions, one on top of the other, represent the museum's east wing. The lower building with the square floor plan by Mendini is complemented by a spectacular pavilion by Coop Himmelb(l)au. It contrasts sharply with the parts of the museum that were designed by Mendini, characterized by their clear forms. The design by Coop Himmelb(l)au consists of large two-sided steel slabs, which, when not touching, alternate with tempered glass. The slabs seem randomly placed and sometimes extend over the underlying pavilion.

Kunstmuseum

Das Kunstmuseum der Stadt Groningen ist nicht nur ein Ort für die Kunst. Allein das Gebäudeensemble ist schon einen Besuch wert. Das Konzept für die Pavillons entwarf der italienische Designer Alessandro Mendini in Zusammenarbeit mit Michele de Lucchi, Philippe Starck und Coop Himmelb(l)au. Das Museum liegt auf einer Insel mitten im Verbindungskanal der Stadt. Zwei übereinander liegende Pavillons bilden den östlichen Museumsflügel. Der untere Baukörper mit quadratischem Grundriss von Mendini wird durch einen spektakulären Aufbau von Coop Himmelb(l)au ergänzt. Er steht in starkem Kontrast zu den von Mendini entworfenen Bereichen des Museums, die sich durch ihre klaren Formen auszeichnen. Der Entwurf von Coop Himmelb(l)au zeichnet sich durch große doppelwandige Stahlplatten aus, die an den Stellen, an denen sie sich nicht berühren, mit gehärtetem Glas abwechseln. Die Platten stehen kreuz und quer und ragen an einigen Stellen über den darunter liegenden Pavillon hinaus.

艺术博物馆

荷兰格罗宁根市艺术博物馆不仅因为其中的艺术品而著称，其建筑本身也值得专门参观。博物馆规划设计的概念由意大利设计师亚历山德罗·门迪尼提出并与米凯莱·德卢基、菲里普·斯塔克、蓝天组进行了合作。博物馆位于城市运河中部的一个水中岛上，两座穿插叠合的展馆形成了博物馆的东翼。门迪尼负责下部方形展馆的设计，蓝天组则设计了一栋令人惊叹的展馆作为项目的补充，这个部分与前者设计的、外观简洁的展馆形成鲜明对比。蓝天组的设计中有两块巨大的钢结构板，它们好像永远也不愿相交，钢化玻璃穿插布置在建筑中，钢板的组织方式好像很随意，但有时它们又像是其他展馆的某种延续。

A red pier leads from the shore over the water to the exceptional museum building, which towers out of the water like a sculpture.

Ein roter Steg führt vom Ufer über das Wasser zu dem außergewöhnlichen Museumsbau, der wie eine Skulptur aus dem Wasser ragt.

一座跨过水面的红色扶梯折桥，连接河岸与这座不同寻常的建筑，就如水上的雕塑。

Through the seemingly random placement of the steel slabs, exciting glimpses and views onto the surrounding water are possible. In some places, it seems to penetrate the museum's interior.

Durch die scheinbar willkürliche Anordnung der Stahlplatten ergeben sich immer wieder spannende Ein- und Ausblicke auf das umgebende Wasser. An manchen Stellen scheint es bis in die Innenräume vorzudringen.

通过那些看似随意安置的钢板，建筑脱颖而出、成为整体环境中令人兴奋的部分，从某些角度，它们看起来好像穿透了建筑的内部。

德鲁根·梅斯尔建筑师事务所

Ray 1 House

In the attic of an office building in the fourth district of Vienna, architects and married couple Elke Meissl and Roman Delugan have created an unusual abode. Flowing transitions, both inside and outside, define the basic outline of this ultramodern design. A spacious, frameless glazed corner window opens completely, thus extending the living area onto a roof terrace with a protruding water basin. This basin takes on the function of mandatory fall-protection, so the panorama view over Vienna is unencumbered by guardrails.

Haus Ray 1

Auf dem Dach eines Bürohauses im Vierten Wiener Bezirk haben sich das Architektenteam und Ehepaar Elke Meissl und Roman Delugan ein ungewöhnliches Privatdomizil geschaffen. Innen wie Außen bilden fließende Übergänge den Grundentwurf für das ultramoderne Design. Eine großzügige rahmenlose, vollständig aufschiebbare Eckverglasung öffnet den Wohnraum hin zu einer Dachterrasse mit vorgelagertem Wasserbecken, das einen schwellenfreien Panoramablick über Wien ermöglicht. Die Wasserbecken ersetzen die vorgeschrieben Geländer zur Absturzsicherung.

鳐式住宅1号

在维也纳第四区一栋办公楼的楼顶，一对建筑师夫妇艾克·梅斯尔和罗曼·德鲁根设计了一处独特的住所。建筑内部和外部流畅的过渡和处理，阐述了这一超现代的设计。一扇巨大的、无框玻璃角窗贯通到底，使得起居空间延伸至建有出挑水池的屋顶平台，水池起到防跌保护的作用，没有阻挡视线的栏杆，可以使维也纳的美景尽收眼底。

The basin glows diffusely in the night and extends to the rooftop edge of the underlying office building.

Das Wasserbecken leuchtet in der Nacht diffus und grenzt bis an den Dachrand des darunter liegenden Bürohauses.

池水在夜间熠熠闪光，一直延伸到下层办公楼的外沿。

The inner and outer borders seem to dissolve thanks to the large-scale glass front. Water borders directly on the living area.

Durch die großflächige Verglasung scheinen Innen- und Außenraum zu verschmelzen. Das Wasser grenzt direkt an den Wohnraum.

内外边界似乎融合在巨大尺度的玻璃前。水池环绕在生活区域之外。

登顿·科克·马歇尔与罗伯特·欧文建筑师事务所

Webb Bridge

Like a snake out of steel mesh, the *Webb Bridge* slithers along the Yarra River in Melbourne, Australia. It was part of an art project in the Melbourne Docklands and was designed by the artist Robert Owen and the architect. Constructed out of steel beams, it resembles a feather gliding over the river landscape, with its flat curve, large span and completely optimized fish bellied beams, so as to ensure complete lightness. Exclusively for pedestrians and cyclists, this bridge connects the Docklands with new residential complexes. The already existing bridge measured 475 feet (145 meters) and was extended by the curved construction by about 262 feet (80 meters). It offers an exceptional perspective of the Melbourne Docklands architecture. In its attempts to be near the Yarra River, the *Webb Bridge* makes water increasingly tangible.

Webb Bridge

Wie eine Schlange aus Stahlgeflecht windet sich die *Webb Bridge* über den Yarra River in Melbourne. Sie ist Teil eines Kunstprojekts in den Docklands der Stadt und wurde von den Architekten in Zusammenarbeit mit dem Künstler Robert Owen entworfen. Ein aus Stahlträgern konstruierter, in seiner flachen Krümmung und großen Spannweite auf absolute Leichtigkeit hin optimierter Fischbauträger liegt einer Feder gleich über der Flusslandschaft. Die ausschließlich für Fußgänger und Fahrradfahrer zugängliche Brücke verbindet die alten Docks mit den neuen Wohnanlagen. Die 145 Meter lange, bereits existierende Brücke wurde durch die gekrümmte Konstruktion um 80 Meter verlängert. Sie bietet bei der Betrachtung der Hafenarchitektur eine außergewöhnliche Perspektive und durch die angestrebte Annäherung zum Yarra River wird das Element Wasser direkt erlebbar.

韦布桥

韦布桥就像一条外披钢网的蛇蜿蜒在澳大利亚墨尔本的亚拉河上，它是墨尔本杜克角的一个艺术发展项目，由艺术家罗伯特·欧文和建筑师共同设计。桥的外表是钢结构梁，具有平滑的曲线、巨大的跨度和最优化的鱼腹梁式结构，它看起来十分轻盈，就像一片掠过河面的羽毛。这座桥仅供步行和骑行，连接着杜克角和新的居民区。原有的桥长约475英尺（145米），网状弧形构造部分约有262英尺（80米），它赋予了墨尔本独特的景色。韦布桥试图尽可能的贴近水面，这让亚拉河变得仿佛触手可及。

迪勒+斯科菲尔德建筑师事务所

Blur Building

Blur Building was the temporary emblem of Yverdon-les-Bains, Switzerland, for the World Fair in 2002. Both New York architects Elisabeth Diller and Ricardo Scofidio were given the project of conceiving and executing the construction of a cloud. A delicate steel construction with a height of 65 feet (20 meters) and a ground surface of 196 feet (60 meters) by 328 feet (100 meters) was anchored in the Neuchâtel Lake. More than 30,000 high-pressure stainless steel mist nozzles with a nozzle diameter of 0.004 inches (0.12 mm) shot fine mist, which was immediately absorbed by the air. This resulted in fog, in this case, also known as the 'Blur-effect.' The water was pumped 765 yards (700 meters) out of the depths of the lake and transported via a pipeline into the pump room, where it was filtered and given chloride before being pushed through the pumps and into the nozzles. Visitors would reach the island through two fiberglass platforms, where they were then enveloped in misty fog and seemed to disappear.

Die Wolke

Le Nuage, die Wolke, war das temporäre Wahrzeichen von Yverdon-les-Bains anlässlich der Weltausstellung 2002. Die beiden New Yorker Architekten Elisabeth Diller und Ricardo Scofidio erhielten den Auftrag, den Bau der Wolke zu konzipieren und auszuführen. Eine filigrane Stahlkonstruktion von 20 Metern Höhe und einer Grundfläche von 60 x 100 Metern wurde fest im Neuenburgersee verankert. Mehr als 30.000 Edelstahldüsen mit einer Öffnung von 0,12 Millimetern Durchmesser versprühten feine Tröpfchen, die sofort von der Luft aufgenommen wurden. Dadurch entstand der Nebel, im Fachjargon auch Blur-Effekt genannt. Das Wasser wurde in 700 Metern Entfernung aus der Tiefe des Sees gepumpt, mittels einer Pipeline in den Pumpraum geschafft, dort gefiltert und mit Chlor versetzt und dann über Pumpen in die Düsen geschleust. Die Besucher gelangten über zwei Rampen aus Fiberglas zu der Insel. Dort wurden sie vom feinen Nebel eingehüllt und verschwanden in ihm.

云建筑

云建筑是瑞士伊韦尔东莱班 2002 年世博会的临时性标志物。纽约建筑师伊丽莎白 – 迪勒和里卡多・斯科菲尔德一同构思并完成了“云”设计的建造。高 65 英尺（20 米）、占地为 196 英尺（60 米）乘 328 英尺（100 米）的钢构平台非常精致，被锚固在纳沙泰尔湖底。超过 30000 个直径尺寸为 0.004 英寸（0.12 毫米）的不锈钢高压喷头喷出细小的水滴，它们弥漫在空中形成雾——也就是“云”。湖水从 765 码（700 米）深的湖底抽上来的，它们被管道输送至泵房，再经过滤和添加氯气。参观者经过两个玻璃钢材质的平台到达这个半岛后便被神秘的水雾所笼罩，仿佛立刻就消失了。

理查德·海韦尔·埃文斯建筑师事务所

Zil Pasyon Spa Resort

The hotel and spa resort *Zil Pasyon* are on one of the Seychelles Islands, the picturesque Felicité Island, in the middle of the Indian Ocean. London architect Richard Hywel Evans designed the exclusive accommodations gently embedded in the aquatic landscape. Some of the bungalows are on a narrow promontory. This is where the first spectacular pool is situated, directly competing with the turquoise ocean and the exquisite sand beach. Wooden bridges connect the residential islands with more luxurious houses farther out in the water, all sumptuously decorated. These residences all have private pools, sundecks and direct access to the ocean.

Resort Zil Pasyon

Das Hotel und Spa Resort *Zil Pasyon* liegt auf der malerischen Seychelleninsel Felicité mitten im Indischen Ozean. Der Londoner Architekt Richard Hywel Evans entwarf die exklusiven Unterkünfte, die sich sanft in die Wasserlandschaft einfügen. Einige der Bungalows liegen auf einer schmalen Landzunge. Hier befindet sich auch der erste spektakuläre Pool, dessen Wasser in direkter Konkurrenz zum türkisblauen Ozean und dem feinen Sandstrand steht. Holzstege verbinden die Wohninseln mit weiteren, im Meer liegenden Häusern, die besonders luxuriös ausgestattet sind. Sie verfügen über private Pools auf den Sonnendecks und haben einen direkten Zugang zum Meer.

Zil Piasyon Spa度假村

Zil Plasyon Spa度假村位于印度洋塞舌尔群岛之中的一个小岛——风景如画的费利西泰岛。这一奢华的建筑群由伦敦建筑师理查德·海韦尔·埃文斯设计，度假村自然地融入了海滨景观之中。部分房屋建在狭窄隆起的沙滩之上，这里有令人叹为观止的水疗池，它们与湛蓝的海水、细腻洁白的沙滩一起组成了世界上最为出色的疗养地。长长的木桥连接着那些离岸更远的、更豪华的、被精心装饰的水中建筑，它们都配有私人泳池、日光浴平台和可以直接下海的入口。

Guests can enjoy luxurious treatment in the spa and wellness area of the resort. The calming effect of water plays a significant role here as well—large aquariums, embedded in the walls show the amazing underwater world of the Seychelles Islands.

Im Spa Bereich des Resorts können sich die Gäste verwöhnen lassen. Auch hier spielt die beruhigende Wirkung von Wasser eine große Rolle: Riesige, in die Wände eingelassene Aquarien zeigen die einzigartige Unterwasserwelt der Seychellen.

宾客在度假村中的Spa和健身区可以享受豪华的服务,在这里，海水扮演了重要的角色——它令人心神宁静，而镶嵌在墙面的大型水族馆则让人惊叹不已——塞舌尔的海底世界就在眼前。

福斯特·莱比锡建筑师事务所

Floating Homes

Inspiration for the concept of *Floating Homes* arose out of the fascination the architects feel for every kind of water. They were looking in Hamburg, Germany, for affordable pieces of real estate with access to water. As this proved almost impossible to find, they conceived and developed an innovative concept for unused property with access to water—*Floating Homes*. The cubic buildings out of glass, steel and wood follow a new lifestyle and living trend, namely water as a form of living space. All in all, a total of eight different types with 311 to 1,312 square feet (95 to 400 square meters) living space were planned, of which only the Home b-type was built in the *City Sporthafen Hamburg*. The futurist residential buildings are mainly interconnected through swimming piers. Due to the staggered design, unencumbered views over the water are possible. Protected intermediary zones are also feasible; they can be used to moor sailboats or as floating gardens.

Floating Homes

Die Idee zu den „Schwimmenden Häusern“ beruht auf der Faszination, die die Architekten für jede Art von Wasser hegen. Sie suchten in Hamburg nach bezahlbaren Wohnungsgrundstücken mit Wasseranbindung. Da sich dies als fast aussichtsloses Unterfangen herausstellte, entwickelten sie für ungenutzte Grundstücke mit Wasserflächen ein innovatives Konzept: die *Floating Homes*. Die kubischen Bauten aus Glas, Stahl und Holz folgen einem neuen Wohntrend – der Entdeckung des Wassers als Wohnraum. Entstanden sind insgesamt acht verschiedene Typen mit 95 bis 400 Quadratmetern Wohnfläche, von denen bisher das „Home b-type“ im City Sporthafen Hamburg realisiert wurde. Die futuristischen Wohnhäuser sind größtenteils untereinander durch schwimmende Steganlagen verbunden. Aufgrund der versetzten Anordnung werden einerseits offene Blickbeziehungen über das Wasser ermöglicht, andererseits entstehen geschützte Zwischenzonen, die als Segelboothäfen oder schwimmende Gärten genutzt werden können.

浮动家园

“浮动家园”设计概念的产生，是因建筑师被水的魅力所吸引。在德国汉堡市找到临水、并可以负担得起的房产，这已被证明是奢望，因此针对还未被利用的临水区域，设计师构思并发展了一个创新性的概念——浮动家园。这个外表较为方正的建筑由玻璃、钢和木材组成，它遵循了一种新的居住方式和生活倾向，将水作为生活空间中的一种形式。共有 8 种、面积在 311~1312 平方英尺（95~400 平方米）的户型正在规划之中，这其中只有 b 型已经在汉堡体育城建成，这座未来主义的建筑与漂浮式平台相互连接。得益于这一令人惊奇的设计，当从房子向外看时，港口的景色毫无遮挡，居住者的隐私空间也得到了保障，这些房子可以被看做是锚固的“帆船”或“漂浮的花园”。

The living area, kitchen and dining area all have large windows that face the water. Secondary areas, such as the bathroom, toilet and vestibule face the shore or the wharf. The spacious sky deck lets residents enjoy the fresh air in a sheltered area.

Wohnraum, Küche und Essplatz öffnen sich mit großen Fensterflächen zum Wasser hin. Nebenräume, wie Bad, WC und Windfang liegen zum Ufer oder zur Kaikante. Das großzügige Skydeck bietet den Bewohnern einen geschützten Aufenthaltsbereich an der frischen Luft.

起居室，厨房和餐厅都有面向水面的巨大窗户。其他区域，像浴室、卫生间和门厅则冲着堤岸或港湾。在开敞的露天甲板上，居住者可以舒服地呼吸清新的空气。

兄弟-弗里德里克-船厂

Living on Water

House boats or traditional floating homes are usually unused, discarded boats or raft constructions with a house. The work group *Living on Water,* based in Kiel, Germany, follows a completely different concept. Their floating houses differentiate themselves from a house boat in that they have no motor and, thus, remain immobile. They are conceived as units between boats and houses. This is not only visible in the design, but also in the respective technology. *The Living on Water* houses consist of a ship-like, 'intelligent' floating body out of steel and a house-like structure. They unite the maritime competence of the shipyard with architectural competence and contemporary design. New dimensions are created on all levels—esthetically, technologically and ecologically. The presumed opposition of a boat and a house dissolves, merging into a new living and working form.

Living on Water

Hausboote oder schwimmende Häuser sind vor allem ausrangierte Schiffe oder Floß-mit-Haus Konstruktionen. Ganz anders das Konzept der Kieler Arbeitsgemeinschaft *Living on Water.* Ein Schwimmhaus unterscheidet sich unter anderem dadurch von einem Hausboot, dass es keinen Motor hat und folglich unbeweglich ist. Das Schwimmhaus ist als eine Einheit zwischen Schiff und Haus konzipiert. Dies ist nicht nur im Design zu erkennen, sondern spiegelt sich auch in der Technik wider. So besteht das Schwimmhaus aus einem schiffsähnlichen, „intelligenten" Stahlschwimmkörper und einem hausähnlichen Aufbau. Das Kieler Schwimmhaus vereint maritime Kompetenz der Schiffswerft mit architektonischer Kompetenz und modernem Design. Ästhetisch, technisch und ökologisch werden neue Dimensionen erschlossen und vermeintliche Gegensätze von Schiff und Haus werden harmonisch zu einer neuen Wohn- und Arbeitsform vereint.

水上之家

船屋或者传统上的浮动房屋通常是没有的，一般是以废弃的船只或木筏为基础设计为住所。以德国基尔（Kiel）为经营基地的工业集团“水上之家”公司，有着不同的思路，他们建造的漂浮屋和普通的船屋相比，前者没有马达，房屋保持着固定的状态，被定义为介于船与房屋之间的形态。它不仅体现在设计上，也表现在采用的各种技术之中。它们有船的外观，智能化的漂浮体外部是钢材和房屋化的构造形式。他们将船坞的航行能力、建筑的功能与现代的设计集合在一起。在审美、技术、生态各个层面形成了新的标准。船和房屋是相互对立的这种推论被化解，两者融合为一种新的生活和工作方式。

LIVING ON WATER 1

LADY M

These floating houses offer a variety of possibilities—whether as a one-family house, office, restaurant, vacation home or exhibition space, the concept can be used in countless different ways.

Das Schwimmhaus bietet vielfältige Nutzungsmöglichkeiten: Ob als Wohnhaus, Büro, Restaurant, Ferienhaus oder als Ausstellungsraum, das Konzept ermöglicht eine Vielzahl von unterschiedlichen Varianten.

这种漂浮房屋提供了多种使用上的可能性，无论是家庭住所、办公室、酒店、度假屋还是展览空间，这一概念可以适应不同的使用需求。

Exploded House Project

Bodrum's history in southwestern Turkey can be traced back to antique times. In order to overcome the limits of this heritage and develop new forms of flexible architecture to be integrated in the landscape, GAD, Global Architectural Development, designed a unique composition of functionally separated parts of a residential house, where rain water plays a crucial role. All windows can sink into the ground, so that cool sea air can breeze through the building. Pools on the roof of the vestibule collect rainwater, which helps cool down the building. Water cascades from one pool into another and lowers the room temperature naturally. This represents an enormous advantage in the hot climate. GAD's concept is a new interpretation of traditional building methods. With the water reflection on the roofs, it offers a fascinating contrast to Bodrum's sloping landscape and can be admired from afar.

Exploded House Project

Die Geschichte Bodrums im Südwesten der Türkei lässt sich bis zur Antike zurückverfolgen. Um dieses Erbe zu überwinden und um neue Formen einer flexiblen und in die Landschaft integrierten Architektur zu entwickeln, entwarf GAD, Global Architectural Development, eine eigenwillige Komposition funktional getrennter Wohnhäuser, bei der Regenwasser eine verbindende Rolle spielt. Alle Fenster können in den Boden versenkt werden, so dass kühle Seeluft in das Gebäude strömt. Zusätzliche Kühlung bringen die Regenwasserbecken auf dem Dach des Vestibüls. Das Wasser fließt von einem Reservoir in das andere und senkt so auf natürliche Weise die Raumtemperatur. Ein unschätzbarer Vorteil in dem heißen Klima. Das Konzept von GAD ist eine neue Interpretation traditioneller Bauweisen. Mit seinen Wasserspiegeln auf den Dächern bildet das Wohnhaus einen faszinierenden Kontrast in der hügeligen Landschaft von Bodrum und ist schon von weitem zu sehen.

分离住宅项目

土耳其西南部博德鲁姆的历史可追溯到久远的古代。为了克服传统建筑形式的局限，发展适应性更强的、与环境相协调的新建筑形式，GAD 全球建设发展公司提出了一种独特的构想，房屋具备了新的衍生功能，雨水起到了重要的作用。建筑外窗全部采用一种可局部沉入地面的设计，这有助于凉爽的海风在建筑中对流。为了进一步帮助建筑降温，顶部的水池被用做雨水收集，水从顶部的水池逐级跌落到低处，在这一过程中以自然的方式降低室内温度。GAD 的概念重新诠释了传统的建筑方法，在炎热的气候中有明显的优势。屋顶的池水波光粼粼，远远就能看到，这一景象在博得鲁姆的山地景观中令人陶醉。

弗兰克·盖里

Guggenheim Museum

With its unique architecture, the Guggenheim Museum has become a trademark for contemporary museum construction across the world and not just in Bilbao, Spain, where it is situated. Initially, the futurist building was alien in this city marked by industry and abandoned factories. Yet, ultimately, it had a healing effect on Bilbao. Experts speak of the 'Guggenheim-effect.' Rusting abandoned shipyards have been replaced by cafés and green squares, encouraging visitors to stay and relax. Fashionable boutiques and exclusive hotels have sprung up. The Guggenheim Museum lies on the Nervión River which, formerly polluted through the industry, has now become a popular tourist attraction for pedestrians and cyclists alike.

Guggenheim Museum

Das Guggenheim Museum ist durch seine einzigartige Architektur auch weit über Bilbao hinaus zu einem Wahrzeichen zeitgenössischen Museumsbaus geworden. Anfangs war das futuristische Bauwerk in der von Industrie und stillgelegten Fabriken geprägten Stadt ein Fremdkörper. Es sollte für Bilbao jedoch eine heilende Wirkung haben. Experten sprechen von einem „Guggenheim-Effekt". Wo einst stillgelegte Werften verrosteten, laden jetzt Grünanlagen und Cafés zum Verweilen ein. Bilbao erhielt modische Boutiquen und exklusive Hotels. Der Fluss Nervión, an dessen Ufer das Guggenheim Museum liegt, ist kein stinkendes Industriegewässer mehr, sondern ein beliebtes Ausflugsziel für Spaziergänger und Radfahrer.

古根海姆博物馆

古根海姆博物馆的建筑设计十分独特，它不仅是西班牙毕尔巴鄂的标志，也是全球当代博物馆建筑的里程碑。最初，这座未来主义建筑在这个以工业和废弃工厂而闻名的城市中是一个异类。但最终，它促使毕尔巴鄂浴火重生，专家们开始谈论“古根海姆效应”，锈迹斑斑的荒废船坞被咖啡馆和绿色的广场所替代，让参观者逗留和在此游览放松。流行时装店和高档酒店雨后春笋般出现。古根海姆所在的内尔韦恩河，之前被污染，现如今则成了步行者和自行车爱好者向往的目的地。

Museoa

The shiny silvery titanium shell envelops the museum with its sea of wave movements. Through its form, this façade not only evokes water, the city's vital element, but also reflects the surrounding water of the Nervión River.

Die silbern glänzende Titanhaut umschließt das Museum wie eine riesige Wellenlandschaft. Die Fassade bildet nicht nur das für die Stadt lebenswichtige Element Wasser durch seine Form ab, sie spiegelt und reflektiert auch das umgebende Wasser des Nervión.

如同海浪般起伏，银光闪闪的钛材外壳覆盖着博物馆，建筑的外观不仅唤起人们对水这一生气勃勃的重要城市元素的感触，也在反射着内尔韦恩河的河水。

gmp 建筑师事务所

Liquidrom

The *Tempodrom* in Berlin, Germany, and its two arenas have been around since 1980. Now, a further attraction has been added—the *Liquidrom.* Meinhard von Gerkan, who was originally responsible for the building complex, has now created a spaciously planned salt-water pool and relaxation area. A water basin of 42 feet (13 meters) in diameter is vaulted by a domed concrete shell, with a top light in its zenith. This concert room is the core of the complex, where live music is played for special events. Computer-controlled light shows, underwater speakers and sound columns set an almost magical mood. At a relaxing temperature of 96° F (36°C), visitors float on the surface of the saltwater, virtually weightless. In addition to the circular salt-water pool, the *Liquidrom* also features several saunas and steam baths, an open-air hot water pool, named *Onsen,* in reference to the Japanese tradition, as well as a bar and a restaurant.

Liquidrom

Das *Tempodrom* mit seinen beiden Veranstaltungsräumen in Berlin besteht bereits seit 1980. Nun wurde es um eine Attraktion ergänzt: das *Liquidrom.* Der schon für das *Tempodrom* verantwortliche Architekt Meinhard von Gerkan schuf ein großzügig angelegtes Erholungs- und Solebad. Ein Wasserbecken von 13 Metern Durchmesser wird von einer kuppelförmigen Betonschale mit einem Oberlicht überspannt. Dieser Konzertsaal ist das Herzstück, in dem zu besonderen Anlässen Livemusik gespielt wird. Computergesteuerte Lichtinstallationen, Unterwasserlautsprecher und Klangsäulen schaffen eine stimmungsvolle Atmosphäre. Bei einer angenehmen Temperatur von 36°C schwebt der Besucher nahezu schwerelos auf dem Solewasser. Neben dem kreisrunden Solebecken verfügt das *Liquidrom* auch über mehrere Saunen und Dampfbäder, ein nach japanischer Tradition „Onsen“ genanntes Heißwasser-Freiluftbecken sowie eine Bar und ein Restaurant.

水疗中心

德国柏林的特鲁姆普和它的两个场馆建成于1980年，现在，它增加了一个更富吸引力的场所——水疗中心，作为特鲁姆普最初的负责人，建筑师迈因哈德·冯·格康又设计了一个大型的盐浴池以及休闲区域。水池的直径为42英尺（13米），上方是半圆形的混凝土穹窿顶盖，灯光布置在建筑顶端。这个盐浴池是建筑的核心，它配有实时音乐播放系统，在某些特别时段，通过计算机控制的灯光秀、水底扬声器和音响会营造出一种变幻无常的氛围。温度为36℃的池水令人放松，加有盐分的池水，让人飘飘欲仙。除了这个主要的盐水池，水疗中心还包括桑拿、蒸汽浴、一个按照日本习惯取名为“温泉”（Onsen）的露天热水池，此外，还有一个酒吧和一家餐厅。

尼古拉斯·格里姆肖及其合伙人建筑师事务所

New Royal Bath

Even the Romans enjoyed bathing in the famous hot natural mineral springs in the English city, Bath. For the millennium, five buildings of the baths complex, all listed on the historical register, were completely restored and reopened. A new building, the *New Royal Bath,* by Nicholas Grimshaw, was added. Grimshaw opted for clear and simple lines for the new building, which carefully integrates itself into the complex. The *New Royal Bath* features thermal baths, a gym and massage rooms. The building is designed as a 'house in a house.' A glass shell, which is molded to the course of the road, creates the framework for the three-storied natural stone construction, which incorporates the neighboring bath house from 1775, the *Hot Bath.* In the center of the building, a freely shaped pool is situated, which can be seen from various vantage points. Another pool with a breathtaking panorama view is on the roof of the building.

New Royal Bath

Schon die Römer badeten in den berühmten Thermalquellen der englischen Stadt Bath. Zum Jahrtausendwechsel wurden fünf unter Denkmalschutz stehende Gebäude des Bäderkomplexes renoviert und durch einen weiteren Bau, das *New Royal Bath,* von Nicholas Grimshaw ergänzt. Grimshaw wählte für den Neubau eine klare und einfache Formensprache, die sich behutsam in das Ensemble einfügt. Im *New Royal Bath* befinden sich Thermalbäder, sportliche Einrichtungen und Massageräume. Das Gebäude ist als „Haus im Haus" konzipiert. Eine Glashülle, die sich dem Straßenverlauf anpasst, bildet den Rahmen für einen dreigeschossigen Natursteinbau, der das angrenzende Badehaus von 1775, das *Hot Bath,* einbezieht. In der Mitte des Gebäudes befindet sich ein frei geformter Pool, der von den verschiedenen Ebenen einsehbar ist. Ein weiteres Becken mit einem atemberaubenden Panoramablick befindet sich auf dem Dach des Hauses.

新皇家浴室

即便是罗马人也喜欢在英格兰巴斯城的天然矿石温泉泡澡。为了庆祝 2000 年千禧年，由 5 栋建筑组成的浴室，均在历史原址上完成了重建并再次开张。新增的建筑之一 —— “新皇家浴室”，由尼古拉斯・格里姆肖公司设计，设计师采用了简练的手法，稳妥地将其融入整体的建筑群中。新皇家浴室包括有温泉浴室、健身房和按摩室。建筑面向道路，就像“房中房”一样，玻璃材质的外壳形成了这栋三层石材建筑的外观构架，与紧邻的建于 1775 年的浴室建筑—— “热屋”相呼应。在建筑的中心部位，一个形状自由的水池从很多地方都可以被看到。另外一处水池建在建筑的屋顶，有着令人惊叹的全景视野。

Turquoise water offers immediate relaxation and highlights the deliberately created opposites —light and shadow, transparency and borders, highly geometrical and softly flowing forms contrast delightfully, setting an exciting, invigorating mood.

Das türkisblaue Wasser lädt zum Entspannen ein und lässt den Blick auf die bewusst inszenierten Gegensätze zu: Licht und Schatten, Transparenz und Geschlossenheit, strenge geometrische und weich fließende Formen wechseln sich ab und bauen so ein spannungsvolles Bild auf.

碧绿的池水令人倍感轻松，有意识的重点照明形成了光影的反差，空间既通透又有边界感，严谨的几何关系和流动的形式之间的对比令人愉悦，它们共同塑造了令人兴奋的清爽氛围。

格鲁普综合馆

Aua extrema

Aua extrema translates from Romanian to mean 'varied, extreme water.' This is also how this extraordinary exhibit was presented during the World Fair in 2002, in Neuchâtel, Switzerland. On a platform in the middle of the lake, the exhibition areas were sheltered by large, plate-shaped roofs. Water was a ubiquitous subject in the interior of the fair, as well. Around 2,500 dripping water hoses created a water wall. Natural and artificial worlds contrasted climatically with the Frost House, where the cold, a hot floor and hot air (82°F, 28°C) helped create beautiful frost images, just like those on the windows of a little cabin in the mountains during wintertime.

Aua extrema

Aua extrema bedeutet im räto-romanischen „vielfältiges, extremes Wasser". Und so präsentierte sich auch diese außergewöhnliche Ausstellung anlässlich der Expo 2002 in Neuenburg. Auf einer Plattform mitten im See überspannten große tellerförmige Dächer die Ausstellungsflächen. Und auch im Inneren war Wasser das allumfassende Thema. Rund 2.500 tropfende Wasserschläuche bildeten einen Wasservorhang. Der Kontrast von Natur und Künstlichkeit erreichte im phantastischen Eisblumenhaus seinen Höhepunkt: Heißer Boden, heiße Luft (28°C) – und rundum echte Eisblumen wie an den Fenstern eines winterlichen Berghauses!

多变的水

"Aua extrema"一词源自罗马尼亚语，意为"多变，极端的水"，是这一特别的博览会在2002年瑞士纳沙泰尔世博会期间所展示的。在湖中水上平台的展览区被巨大的碟形屋顶所遮盖。这届博览会中，水是无处不在的主题。约2500套滴水软管形成了一道水幕，自然和人工环境通过气候的反差形成结霜屋——低温、热地板和热气（28℃）共同形成美丽的霜花，真的就像冬天里山区小屋的窗户一样！

la rete delle reti
la rait da

Nordsee
Rotterdam
Säntis 2501 m
Tödi 3614 m
das netz der netze le réseau des réseaux the web of webs la rete delle reti la rait da las raits

The large lake of *Aua extrema* intended to highlight this vital element, water. Everything was thus literally placed under water—the floor was lightly misted with water, water dripped down from the roof and sprayed out from the 'walls.'

Der große See von *Aua extrema* sollte den Blick auf dieses lebenswichtige Gut öffnen. So stand hier buchstäblich alles unter Wasser: Der Boden wurde benetzt, vom Dach tropfte es und von den „Wänden" spritzte es.

Aua extrema展览周边的大片湖水试图唤起人们对"水"这一重要对象的关注，所有的设计也都围绕水进行构思——水笼罩着地面，水从屋顶滴落、从水"墙"喷溅。

格维兹门迪尔·琼森

NORVEG Museum and Cultural Center

The small seaside town Rørvik is situated far up in the north of Norway, in the Vikna archipelago. What it lacks in size, it makes up for in traditions. People have lived here on the shore for centuries. Today, Rørvik has primarily become the center of administration and crossroad for the postal ship line *Hurtigruten.* It seems as though this town was predestined to have a museum and cultural center such as Norveg, which is dedicated to coastal life. The new building by Gudmundur Jonsson built along the shoreline opened in 2004. The striking, sail-like rooftops subtly evoke the maritime location and old Viking traditions.

NORVEG Museum und Kulturzentrum

Das Hafenstädtchen Rørvik liegt hoch im Norden Norwegens im Schärengebiet Vikna. Es ist nicht groß, aber traditionsreich. Seit Jahrtausenden leben in dieser Landschaft Menschen von und mit der Küste. Heute ist Rørvik vor allem Verwaltungsort und Kreuzungspunkt der Postschifflinie „Hurtigruten". So ist der Ort geradezu prädestiniert, um mit dem Norveg ein Museum über das Küstenleben und ein Kulturzentrum zu etablieren. Der Neubau von Gudmundur Jonsson liegt direkt am Wasser und wurde 2004 eröffnet. Die markanten, segelartigen Dachflächen verweisen auf die maritime Lage und die alte Wikinger-Tradition.

诺威尔博物馆和文化中心

海滨小镇勒尔维克远在挪威极北部的维克纳（Vikna）列岛。虽然规模很小，但它一直承袭着传统，当地人已经在此居住了数个世纪。现在，勒尔维克成了行政中心和游轮（海峡游）航线上的重要站点，看起来小镇就注定要修建一个诺威尔这样用以纪念海岸生活的文化中心和博物馆。新建筑紧靠海岸，由格维兹门迪尔琼森设计，在2004年开放。光滑的像帆一样的屋盖，隐晦得让人意识到航海和古老的维京传统。

KAI FORBUDT

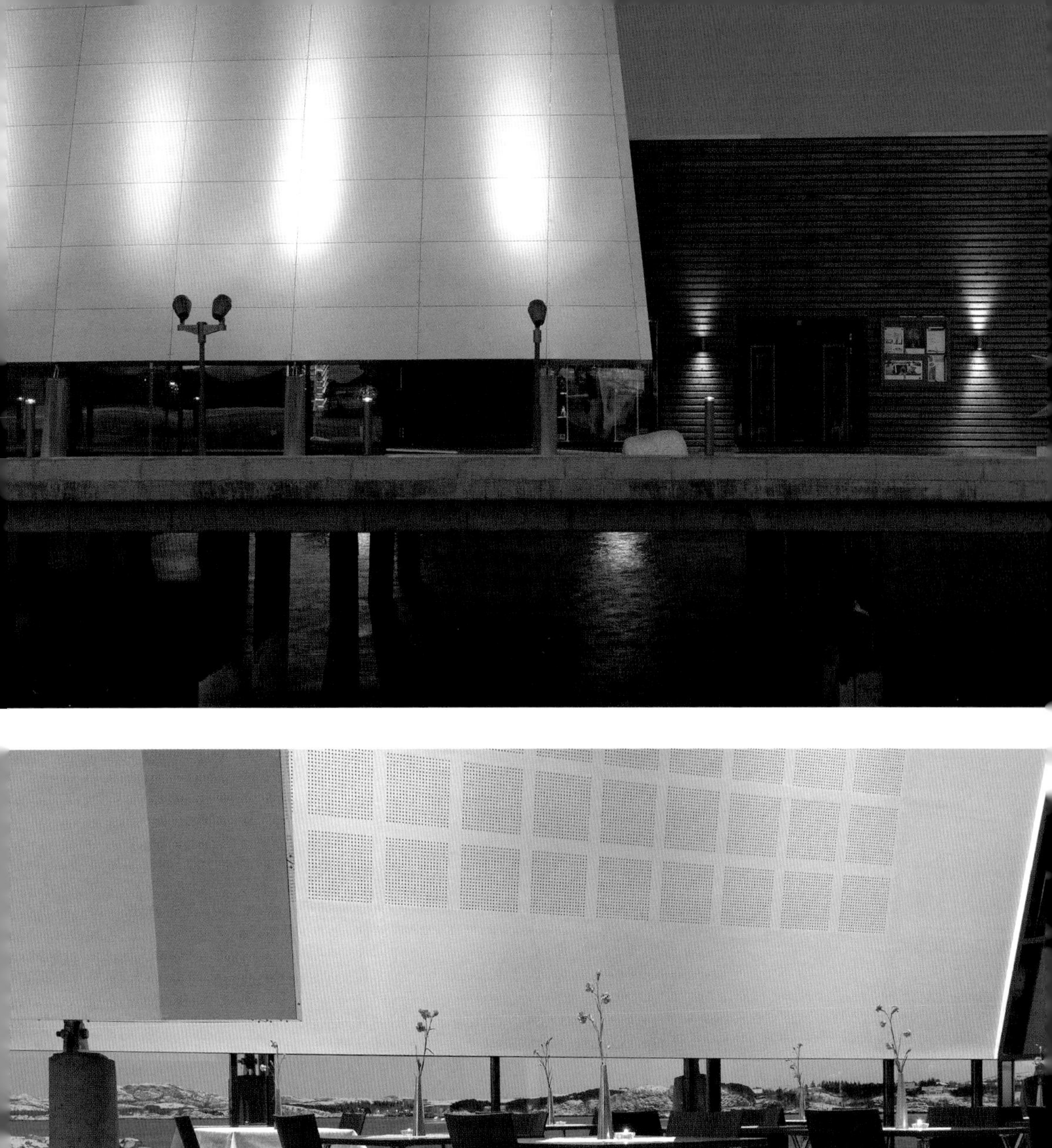

The striking rectangular roofs evoke sails blowing in the wind and give the museum building the silhouette of a Viking ship setting sail.

Die markanten, rechteckigen Dächer erinnern an vom Wind geblähte Segel und verleihen dem Museumsbau die Silhouette eines in See stechenden Wikingerschiffes.

光滑的长方形屋盖如同风中飘扬的船帆，博物馆如同扬帆航行的维京船。

扎哈·哈迪德

BMW Plant

Many requirements needed to be taken into account in the design of the new headquarters of BMW in Leipzig, Germany. On the one hand, this building needed to connect the production halls of the car body construction, paint shop and assembly. On the other, it also needed to function as the main entrance to the factory and to comprise office and communication areas, the cafeteria and various laboratories and workshops. Another requirement was the design of an open and communicative space, where the BMW automobile production process could become more transparent for visitors and employees alike. Zaha Hadid met the clients' requirements with flair when creating his design of a building, where shiny silver façade elements alternate with large-sized window fronts. The façade glitters in quiet water, and the reflection offers a second visual level to the design.

Zentralgebäude der BMW Niederlassung

Die Anforderungen an das neue Zentralgebäude der BMW Niederlassung in Leipzig waren hoch. Zum einen sollte es die Produktionshallen des Karosseriebaus, der Lackiererei und der Montage miteinander verbinden. Zum anderen ist es der Haupteingang zum Werk und beherbergt Büro- und Kommunikationsflächen, das Betriebsrestaurant und verschiedene Labore und Werkstätten. Eine weitere Anforderung war die Realisierung eines offenen und kommunikativen Raums, in dem die BMW Automobilproduktion für Besucher und Mitarbeiter gleichermaßen transparent wird. Zaha Hadid hat die Wünsche der Auftraggeber eindrucksvoll in ihrem Bau aus silbern glänzenden Fassadenelementen, die sich mit großzügigen Fensterflächen abwechseln, verwirklicht. Das Zentralgebäude liegt an einem künstlich angelegten Wasserbecken. Die Fassade schimmert im ruhigen Wasser und verleiht dem Entwurf eine zweite Ebene.

宝马工厂

针对宝马在德国莱比锡的新总部（中心建筑）的设计要求很高。首先，建筑要将车身生产厂房、喷涂车间和装配车间进行串联并结合在一起；其次，它要具备工厂主入口区的功能，要有办公区、会议区、咖啡室、各种实验室和操作间；再有，需设计一个开放、交流的空间，在这里，参观者要和雇员一样，能清楚地了解宝马汽车的生产过程。扎哈·哈迪德以她天才的设计满足了业主的愿望，中心建筑在一个人工水池旁边，建筑外观由相互穿插的银色材质和大尺度玻璃窗构成。平静的水面之中，明亮的倒影和建筑本身相映成趣。

库尔特·霍夫曼

Hotel Palafitte

On the shore of Lake Neuchâtel, a part of the unique Three Lakes region in Switzerland, an extraordinary concept was designed. The entire complex of this luxury design hotel is built on stakes and is either right next to the shore or in the middle of the lake. An ingenious pathway of wooden piers leads out onto the lake and connects the wooden huts with the shore and the restaurant. All pavilions have a terrace and many have direct access to the water. Each pavilion has a large room with a work area, a bathroom with lake view and an outside terrace.

Hotel Palafitte

Am Ufer des Neuenburger Sees, der zur landschaftlich einmaligen Drei-Seen-Region zählt, wurde ein ungewöhnliches Konzept realisiert. Die gesamte Anlage des Designhotels ist auf Pfählen gebaut und steht entweder unmittelbar am Ufer oder mitten im See. Ein ausgeklügeltes Wegenetz aus Holzstegen führt über den See und verbindet die Holzhütten mit dem Ufer und dem Restaurant. Alle Pavillons verfügen über eine Terrasse, viele von ihnen über einen direkten Zugang zum Wasser. Jeder Pavillon umfasst ein großes Zimmer mit einem Arbeitsbereich, ein Badezimmer mit Blick auf den See und eine Außenterrasse.

湖上旅舍

纳沙泰尔湖属于独特的瑞士三大湖区，这里有间构思奇特的旅店。设计奢华的建筑全部通过桩基建在靠近湖岸的水中。一座精巧的木制栈桥通向湖面，连接木屋、湖岸和餐厅。所有房舍都带有平台，很多还设有直接下水的出口。每套房都有包括工作区的大房间、可以眺望湖景的浴室和户外平台。

Guests have an amazing view of the unique Swiss lake landscape from the bungalow terraces. An extended roof and shades on the side of the bungalows offer protection from intense sunrays and privacy from curious neighbors.

Von den Terrassen der Bungalows hat man einen wunderschönen Ausblick auf die einzigartige Schweizer Seenlandschaft. Ein vorspringendes Dach und eine Seitenblende bieten Schutz vor zu starker Sonneneinstrahlung oder den neugierigen Blicken der Nachbarn.

客人在木平台上可以尽揽令人陶醉的瑞士湖区美景，探出的屋檐和隔板起到遮阳和保护客人隐私的作用。

JSK 建筑师事务所

Medienhafen

Transformation of unused port buildings to new office and living quarters, integrated into the city, is something which countless cities will need to do in the future. One of the most successful and outstanding examples of this kind of urban development is the transformation of the *Medienhafen* in Düsseldorf, Germany. The variety of the new uses and the quality in architecture and design of the resulting buildings are visible in the innovative attic restorations, in the modern one-family houses and in the bridge constructions. The historical atmosphere is kept and integrated in a new city district, which is directly defined by the water, with its careful architecture and expressive design of open spaces. Architecture thus represents a crucial element for the desired invigoration of the *Medienhafen* in Düsseldorf, and the urban construction required therefore.

Medienhafen

Die Umwandlung von ungenutzten Hafengebieten in neue, in die Stadt integrierte Wohn- oder Büroquartiere ist eine Aufgabe, die in zahlreichen Städten ansteht. Die Verwandlung des Düsseldorfer Medienhafens gehört zu den herausragenden und erfolgreichen Beispielen einer modernen Stadtentwicklung. Die Vielfalt der neuen Nutzungen sowie die architektonische und gestalterische Qualität der Bauwerke zeigt sich in den innovativen Speicherumbauten, den modernen Wohnhäusern und den Brückenbauten. Das historische Hafenambiente wurde erhalten und in ein neues Stadtviertel integriert, das durch eine sorgfältige Architektur und durch eine ausdrucksstarke Freiflächengestaltung direkt am Wasser geprägt ist. So stellt die Architektur einen maßgebenden Bestandteil der städtebaulich gewünschten Belebung des Düsseldorfer Medienhafens dar.

媒体港

将废弃的港口建筑转化为新式的办公室和生活广场并自然地融入周边环境，是无数城市在未来必将要做的工作。德国杜塞尔多夫的媒体港，就是这类城市改造中最为成功和突出的例子。在创造性的顶楼重建中，在现代公寓建筑上，在桥梁的建设中，多样化的新用途、优良的建筑品质和预期中的设计显而易见。传统的海港环境特征得以保留并融入了新的城市，这个过程中，水起到了重要的作用。通过精心的建设和具有表现力的开放空间设计，建筑学的确表达了杜塞尔多夫媒体港所渴望的新能量中最为关键的部分，而这也是所有城市所需要的。

Lido
Lido
Lido

Just a few years ago the old main port in Düsseldorf inspired little more than bleakness. Now, more than 500 companies have found a new home, where they can enjoy the unique location and benefit from the *Medienhafen's* new image. The port was converted into an area, wherein media, fashion, art and advertisements are located.

Wo noch vor einigen Jahren die Tristesse des alten Düsseldorfer Haupthafens herrschte, haben mehr als 500 Unternehmen Quartier bezogen, nutzen die einzigartige Lage, profitieren vom Image des Medienhafens. Der Hafen wurde umgestaltet zu einem Areal, in dem Medien, Mode, Design, Kunst und Werbung ihre besondere Adresse gefunden haben.

就在几年之前，这个老旧的港口还前景黯淡，了无生气。现在，已有超过500家公司搬迁至此，他们喜欢这里独特的地理环境并受益于媒体港的新形象。港口成功转化成为一个传媒、时尚、艺术和广告行业的聚集地。

汤姆斯·克鲁普建筑师事务所

Universsum Science Center

Thomas Klumpp's design of the *Universum Science Center* in Bremen resembles a UFO, a clam or a whale. For the curved form with an outer skin of around 40,000 stainless steel tiles gives rise to many associations. Like a fish with shiny scales, the museum is surrounded by a lake, which was explicitly created for this purpose. Piers and platforms lead visitors into the interior. An impressive permanent exhibition displays pieces dedicated to one of the three categories Expedition Mankind, Expedition Earth and Expedition Cosmos. In direct proximity to the *Universum Science Center,* further attractive sights have sprung up—the so-called discovery park, an open-air scientific adventure landscape with several interactive stations, landscape elements and an 88-feet (27 meter) high tower of the skies with a cubic viewing post.

Universum Science Center

Das von Thomas Klumpp gestaltete *Universum Science Center* in Bremen sieht wie ein Ufo, eine Muschel oder ein Wal aus. Die geschwungene Form mit einer Außenhaut aus etwa 40.000 Edelstahlschindeln lässt viele Assoziationen zu. Wie ein Fisch mit silbern glänzenden Schuppen liegt das Museum im eigens angelegten See. Über Stege und Rampen gelangt der Besucher ins Innere. Hier zeigt eine beeindruckende Dauerausstellung Exponate zu den Themengebieten Expedition Mensch, Expedition Erde und Expedition Kosmos.
In unmittelbarer Umgebung des Bremer *Universum Science Centers* sind noch weitere attraktive Sehenswürdigkeiten entstanden: Der so genannte Entdecker-Park, eine wissenschaftliche Erlebnislandschaft unter freiem Himmel mit vielen Mitmach-Stationen, Landschaftselementen und einem 27 Meter hohen Turm der Lüfte und einer würfelförmigen Schaubox.

宇宙科学中心

托马斯·克鲁姆普设计的不来梅宇宙科学中心是一栋像UFO、蚌或鲸鱼一样的建筑。由约40000块冲压成形的不锈钢片构成的弧形外表面令人产生很多联想。博物馆建在湖水中，好似一条披着闪光鳞片的鱼，水池是为此专门修建的。水上的平台和坡道引导着参观者进入博物馆的室内，专门展现人类探险、地球探险和宇宙探险的永久收藏展览令人印象深刻。紧邻宇宙科学中心更有吸引力的景点雨后春笋般出现——所谓的探索公园，它是由一些互动装置、各种景观和88英尺（27米）高塔上的方形观景台组成的户外科学探险场所。

马尔西奥·科根

House Mirindiba

This one-family house in São Paulo, Brazil, impresses with the select, luxurious materials used, the well-chosen color compositions and the architect's minimalist language. The entrance area consists of a small atrium, which connects the individual areas. On the ground floor, a path connects the kitchen, dining and living area, while a stairway leads up to the bedrooms and the secluded roof terrace. The spacious living room opens up to an inner courtyard. Two floor-to-ceiling window bands connect both courtyards and allow amazing views. The annexed exterior was designed in accordance to the rest of the house's layout, using clear geometrical structures. Above all, the extended pool attracts immediate attention—it creates an elegant contrast to the untreated stones of the outer walls.

Haus Mirindiba

Dieses Wohnhaus in São Paulo besticht durch die Auswahl edler Materialien, die stimmige Farbkomposition und die minimalistische Architektursprache. Ein kleines Atrium bildet den Eingang und verbindet die einzelnen Bereiche. Ein Weg führt im Erdgeschoss zur Küche und zu den Wohn- und Essbereichen, eine Treppe führt nach oben zu den Schlafräumen und der intimen Dachterrasse. Das großzügige Wohnzimmer öffnet sich zu einem Innenhof. Zwei Fensterbänder, die zwischen Boden und Decke laufen, verbinden die beiden Höfe miteinander und lassen Durchblicke frei. Der angegliederte Außenbereich wurde passend zum Rest des Hauses in klaren geometrischen Strukturen angelegt. Vor allem der lang gezogene Pool lenkt alle Blicke auf sich: Er bildet einen eleganten Kontrast zu den rau belassenen Steinen der Außenwände.

Mirindiba住宅

这个家庭住宅位于巴西圣保罗，名贵材质的运用、色彩的精选搭配和建筑师的极简主义手法，令人印象深刻。入口是一个小的天井，它连接着其他各个独立区域。一层有条小通道连接厨房、餐厅和起居室，楼梯通往上面的卧房和僻静的屋顶天台。宽阔的起居室朝着内部的庭院敞开，两排通顶的落地窗贯通两个庭院，形成让人兴奋的全景视线。对简约几何式构成的运用，少就是多的手法使得建筑让人回味。尤其是长长的水池——它与围墙的粗糙毛石之间形成优雅的对比，会立刻吸引人的注意。

From the dining area and living area, the room-high glass windows give an amazing view of the blue-green shimmering pool. Exterior and interior seem to merge seamlessly.

Die raumhohen Glasscheiben geben im Ess- und Wohnbereich den Blick frei auf das blaugrün schimmernde Wasserbecken. Außen- und Innenraum scheinen miteinander zu verschmelzen.

从餐厅和起居室高大的玻璃窗可以一览无遗地看到外面的盈盈碧水，室内外完美的融为一体。

隈研吾建筑都市设计事务所

Baisouin Temple

This basic concept of a Buddhist temple in Tokyo aimed to counter-balance the dramatic surroundings with simple and clear architecture. A delicate wood-glass cube, placed on a retracted concrete column, lets the building seem to float airily over the urban landscape. The impression of flying is reinforced by the thin flat roof, which extends over the building itself. As is often the case, Kengo Kuma has designed a building, whose architecture merges with the landscape. The clearness of the building is additionally emphasized by the strict harmony of all details and their exact execution. A water level elegantly covers the protruding surface, expands so as to then connect with the environment. Magnificently reflected, the temple seems to swim. The building becomes one with the atmosphere, where space is limitless and boundaries dissolve.

Baisouin Tempel

Die Grundidee des buddhistischen Tempels in Tokio war es, der Theatralik des Ortes eine einfache und klare Architektur entgegen zu setzen. Ein filigraner Holz-Glas-Kubus, der auf einem zurückgesetzten Betonsockel aufsitzt, lässt das Gebäude mit einer gewissen Leichtigkeit über der Stadt schweben. Dieser Eindruck wird durch das rundum weit überstehende und dünne Flachdach verstärkt. Kengo Kuma entwarf ein Gebäude, dessen Architektur mit der Umgebung verschmilzt. Die Klarheit des Baukörpers wird durch die strenge Abstimmung der einzelnen Details und ihrer exakten Ausführung zusätzlich unterstrichen. Eine Wasserschicht bedeckt anmutig die vorgelagerte Fläche, dehnt sich nach Außen aus und verbindet sich mit der Umgebung. Auf diesem Spiegel scheint der Tempel zu schwimmen. Somit wird das Gebäude zu einem einheitlichen Ambiente, in dem es keine Zergliederung der Räume gibt und wo die Grenzen verschwinden.

梅窗院

针对位于东京的这个佛教寺院，设计的基本概念是以简单而清晰的建筑去平衡周边纷杂的环境。由木材和玻璃构成的精致方盒子建筑由下部的混凝土柱支撑，姿态轻盈的漂浮在城市之上，伸展而出的纤薄平顶加强了这种灵动的感觉。就像他之前的那些设计，隈研吾完成了一个与外部环境相融合的项目。建筑的清晰被一再加强，所有严谨一致的细节和精确到位的安排强化了建筑的简洁明了。平静的水面优雅地覆盖在前部的屋面上，向外延伸着，仿佛和天空相连。令人恍惚的水面倒影，如同漂浮的庙宇，建筑和周边的环境融为一体，突破了空间和边界的约束。

Water/Glass Villa

With his design *Water/Glass Villa,* Kengo Kuma remains true to his principles of embedding buildings in natural surroundings and giving them a new dimension. He thus continues the architectural method that fascinated Bruno Taut in the 1930s. Kuma, however, takes an unusual approach, using glass as the main material in his design and water as the element to which he creates a connection. The fascinating end result, the *Water/Glass Villa,* is in Atami, south of Tokyo. Conceived as a guest house for the Bandai group, it was built in the same neighborhood of Taut's Hyuga Villa. Even the location, a steep cliff overlooking the Pacific, is defined by water and becomes a part of the building through the use of transparent materials. The glass floors of the upper levels surpass the glass walls and reach up into the rippling water. Water and glass become unified in the sunlight, merging in the glow of the integrated floor lighting.

Water/Glass Villa

Mit seinem Entwurf *Water/Glass Villa* folgt Kengo Kuma seinem Bestreben, Gebäude in die natürliche Umwelt einzubetten und ihnen eine neue Dimension zu verleihen. Damit führt er zugleich weiter, was auch den Architekten Bruno Taut in den 1930er Jahren faszinierte. Allerdings geht Kuma ungewöhnliche Wege, denn Glas ist das beherrschende Material seines Entwurfs und Wasser ist das Element, zu dem er eine Verbindung herstellt. Die faszinierende *Water/Glass Villa* steht in Atami, südlich von Tokio. Als Gästehaus des Bandai Konzerns wurde sie in der Nachbarschaft von Tauts Villa Hyuga erbaut. Schon die Lage, ein Steilhang mit Blick über den Pazifik, ist vom Wasser geprägt und wird durch die transparenten Materialien zum Bestandteil des Gebäudes. Die Glasböden der oberen Ebene reichen außerhalb der gläsernen Hauswände in eine sanft gekräuselte Wasserfläche hinein. Wasser und Glas verschmelzen im Sonnenlicht oder im Schein der integrierten Bodenbeleuchtung zu einer Einheit.

水–玻璃别墅

通过“水－玻璃”别墅的设计，隈研吾坚守“建筑植入环境理论”的实践并提供了新的诠释。他是如此执着地继续着20世纪30年代令布鲁诺·陶特着迷的建筑思想。不同的是，隈研吾做了更进一步的探索，玻璃是他在设计中使用的主要材料，水作为一种要素，他创造了两者之间的联系。令人陶醉的成果是位于东京南部热海的“水－玻璃”别墅。作为万代集团的迎宾别墅，它和陶特的日向别墅位于同一街道，甚至两者的选址也相同——俯瞰太平洋的险峻峭壁上。建筑被水环绕，由于使用了透明的材料，水和建筑好像成为一体。顶层的玻璃地板穿过玻璃幕墙直通水面，在日光下玻璃和水难分彼此，夜幕降临，它们一起融入整体化地板照明的幽光中。

吕迪格·莱纳

Absberggasse School

The *Absberggasse School* in Vienna, Austria, was built on an open square surrounded by large trees, in the middle of many examples of modern urban architecture of the 1960s. The school building has a comb-like structure, which stimulates constructive interaction with its clear, but not too severe lines. The classrooms are in the secondary wings, which branch off from the long main tract. The two gymnasiums stacked on top of each other are striking, as is the wing for special classes, which sits atop beams and extends over the entrance and the pond. The school yard and the sports field are surrounded by a spacious body of water.

Schule Absberggasse

Die Hauptschule Absberggasse in Wien entstand auf einer freien, von großen Bäumen eingefassten Fläche inmitten von Scheibenhäusern, den Repräsentanten moderner Stadtarchitektur der 1960er Jahre. Das wie ein Kamm strukturierte Schulhaus lädt mit klaren, aber keineswegs strengen Linien zu konstruktiver Interaktion ein. Von dem langen Haupttrakt zweigen Sekundärtrakte ab, in denen die Klassenzimmer liegen. Markant sind die beiden aufeinander gestapelten Sporthallen und der auf Stützen ruhende Trakt der Sonderklassen, der über den Eingang und die Teichanlage weit hinausragt. Auch Pausenhof und Sportplatz sind von der großzügigen Wasserfläche umrahmt.

巴斯伯格斯学校

巴斯伯格斯学校位于奥地利维也纳，建在一个周边是大片树林环绕的地方，周围是很多20世纪60年代的典型现代城市建筑物。学校是梳状外形的平面布局，通过这样简明但并不是很简单的设计来激发学生建设性的互动行为，教室布置在第二排，从长长的主楼分支而出。两个拼合而成的健身房非常引人注目，它们是专门用做特殊课程的，建筑下面有大梁，一直延伸到入口和水池的上方。学校和体育场被大片的水面所环绕。

18 KLASSEN·SPE
PHYSIKCHEMIEUN
BIOLOGIERÄUME·ED
PSYCHAGOGE·PHOT
LABOR·MEHRZWECK
RAUM·SAMMLUN
EINER SCHULE DE
STADT WIEN
MUSIKERZIEHUNG
GARDEROBEN·LEHRE
ZIMMER·BERATU

The imaginative, but in no way fanciful design of the complex, carefully slanted, protruding or narrowed elements and the integration of the pond all prove how architectural components can support pedagogical work.

Die spielerische, aber keineswegs verspielte Gestaltung des Komplexes, behutsam gesetzte Schrägen, Ausbuchtungen, Verengungen und die Integration des Teichs stellen unter Beweis, wie architektonische Komponenten die pädagogische Arbeit unterstützen können.

这是一个充满想象、但并不空泛的建筑设计。精心布局、重点突出、收放自如，还有和谐的水面，这一切都在证明建筑元素是如何适应教学工作需要的。

理查德·迈耶及其合伙人建筑师事务所

Jesolo Lido Village

Only 76 yards (70 meters) from the beach, Richard Meier's *Jesolo Lido Village* was groundbreaking when built along the Adriatic Sea in Italy, surrounded by the large concrete hotels constructed during the building boom of the 1960s. On a ground surface of 98, 425 square feet (30,000 square meters), three-story vacation houses and three towers were planned. Twenty-three apartments and stores represent the center of the complex in the east. A square offers the necessary space and atmosphere to continue the Mediterranean beach life. This part of the complex is complemented by 60 living units along an extended pool. The proximity to the Adriatic Sea is tangible, and the entire complex is marked by Mediterranean light and fresh sea breeze.

Jesolo Lido Village

Nur 70 Meter vom Strand entfernt hat Richard Meier mit dem *Jesolo Lido Village* an der italienischen Adria ein Zeichen gesetzt. Inmitten der im Bauboom der 1960er Jahre entstandenen Bettenburgen wurden auf einer Grundfläche von 30.000 Quadratmetern dreigeschossige Feriendomizile und drei Türme geplant. Im Osten der Anlage bilden 23 Appartements und Geschäfte den Mittelpunkt. Ein Platz bietet Raum und Ambiente für die Fortsetzung des mediterranen Strandlebens. Ergänzt wird dieser Teil der Anlage von 60 Wohneinheiten entlang eines lang gestreckten Pools. Die Nähe der Adria ist überall spürbar und die gesamte Anlage wird von mediterranem Licht und einer frischen Meeresbrise durchflutet.

迪耶索洛丽都度假村

简直就是见缝插针！理查德·迈耶设计的迪耶索洛丽都度假村靠近意大利亚得里亚海，距离海滩只有 76 码（70 米），周围是众多 20 世纪 60 年代开发热期间修建的混凝土结构酒店。在 98425 平方英尺（30000 平方米）的地块上，设计师规划了三栋 3 层高的度假村建筑和三栋塔楼，23 套公寓和商店组成的中心综合楼在建筑群的东侧。一个小广场形成必要的开阔空间和环境，延续着地中海式的海边生活。度假村这一部分沿着长长的水池、由 60 套居住单元组成。临海的感觉非常强烈，度假村到处都是地中海的阳光和清新的海风。

MOS 建筑师事务所

Floating House

The *Floating House* combines the advantages of a ship with those of a house. Thanks to its extreme mobility, it can be easily moved anywhere at any time. This residential boat is in the middle of Lake Huron in Ontario, Canada. Its unique aspect is the proximity to water. The floor is almost level to the water; countless windows and openings increase the impression of walking up and down on the water surface. This aspect and the surrounding landscape create an unforgettable atmosphere.

Floating House

Das *Floating House* verbindet die Vorteile eines Schiffes mit denen eines Hauses. Auf Grund seiner Beweglichkeit kann es zu jeder Zeit an jeden Ort verlegt werden. Das Wohnhaus liegt inmitten des Lake Huron in Ontario, Kanada. Das Einzigartige ist die unmittelbare Nähe zum Wasser. Der Fußboden liegt fast auf demselben Niveau wie die Wasseroberfläche. Zahlreiche Fenster und Öffnungen verstärken zudem den Eindruck direkt auf dem Wasser auf und ab zugehen. Dieser Aspekt und die umliegende Landschaft schaffen eine unvergleichliche Atmosphäre.

浮动住宅

浮动住宅综合了船和住宅两者的优点。得益于特殊的机动性，它可以随时移动到任何地方。这艘居住用船在加拿大安大略省的休伦湖中，其独特之处在于亲水性，它的地板几乎和水面齐平，大量的窗户和开敞空间强化了建筑随着湖水起伏的感觉，和周围景观形成的氛围令人难忘。

The pre-fabricated construction system makes it possible for the wooden construction to be built within 4 weeks. When life and work circumstances change, it can always be adapted accordingly.

Das vorgefertigte Bausystem ermöglichte eine Bauzeit von nur vier Wochen für die Holzkonstruktion. Eine Anpassung an veränderte Lebens- und Arbeitsumstände ist jederzeit möglich.

这座木构建筑采用预制结构系统，在4周之内就可建好。每当生活和工作环境发生变化，它都能做相应的调整以便适应。

MVRDV 事务所

Silodam

Evoking a giant container, this extraordinary one-family house is situated in the ports of Amsterdam. At the end of a pier, this construction is in the proximity of the river Ij and the rather closed construction of the former wood port. The 65 feet (20 m) deep and 10-story high building comprises a total of 142 condominiums and 15 rental apartments. It consists of four building elements that are slightly staggered in height. Different sized and partly strangely shaped apartments were created, such as ateliers, studios, little maisonette apartments and penthouses. The façade cladding varies from wood to brick and each apartment type is colored differently.

Silodam

Wie eine riesige Container-Landschaft wirkt dieses ungewöhnliche Wohnhaus im westlichen Hafengebiet von Amsterdam. Am Ende eines Piers gelegen, befindet sich das Bauwerk im Grenzbereich des Flusses Ij und der eher geschlossenen Bebauung des ehemaligen Holzhafens. Das 20 Meter tiefe und 10 Stockwerke hohe Bauwerk beherbergt insgesamt 142 Eigentumswohnungen und 15 Mietwohnungen. Es besteht aus vier Bauteilen, die in der Höhe leicht versetzt sind. Dadurch entstanden unterschiedlich große und zum Teil ungewöhnlich gestaltete Wohnungen, wie Ateliers, Studios, Maisonettewohnungen und Penthäuser. Die Fassadenverkleidung variiert von Holz bis Backstein und jeder Wohnungstyp besitzt eine andere Farbe.

Silodam公寓

这栋独特的、位于阿姆斯特丹港的公寓楼让人不禁联想到那种巨大的集装箱。它建在一条防波堤的尽头，紧邻艾加河，距离原来的木材码头相当的近。建筑的进深为65英尺（20米），高度为10层，包括142套公寓和15套出租套房。实际上，它由四个部分组成，在高度上稍有不同，形成了不同尺寸和不同形式的公寓住宅，包括工作室、画室、小型复式公寓和顶层阁楼。建筑立面是不同形式的饰面材料，从木材到面砖，并且不同类型的住宅有着不同的色彩。

Built on concrete columns in the water, the building seems as if it could leave for high waters at any given time. And even if this is not possible, the owners can take their boat from their private docks and drive down to the canals of downtown Amsterdam.

Auf Betonpfeilern ins Wasser gestellt, scheint das Gebäude fast jeden Moment in See stechen zu können. Und wenn das auch nicht möglich ist, so haben die Bewohner die Möglichkeit von privaten Anlegestellen direkt mit dem eigenen Boot in die Grachten der Amsterdamer Innenstadt zu fahren.

混凝土立柱上的大楼仿佛随时准备着扬帆远航。尽管可能性不大，但这儿的住户甚至可以从他们的私人码头驾船到阿姆斯特丹中心市区的运河。

努特林斯·雷代克建筑师事务所

The Sphinxes

In the mid-1990s, Neutelings Riedijk designed eight residential blocks on the southern shore of the Gooi-lake in Huizen near Amsterdam. In the meantime, five more buildings were added. Completely built in water, these buildings rise majestically out of the water and resemble sphinxes with their dramatic silhouette. Architects conceived the cross-section of the building like a triangle, whereby one side remains in the water. This exceptional architecture enables the individual residential units to face the open sea to the largest degree possible. The large sun terraces turn to the south towards land. Behind large panorama windows lie the protruding living rooms, of which has each a girth double the width of the apartment. Façades from unpolished raw aluminum elements are supposed to reflect and mirror the gray color of the water and the weather in the Netherlands.

The Sphinxes

Bereits Mitte der neunziger Jahre realisierten Neutelings Riedijk acht Wohnblocks am Südufer des Gooi-Sees in Huizen bei Amsterdam. Inzwischen wurde das Ensemble um fünf komplett im Wasser errichtete Baukörper ergänzt. Sie ragen majestätisch aus dem Wasser und erinnern mit ihren steil abfallenden Silhouetten wie Sphinxe. Die Architekten formten den Querschnitt des Gebäudes wie ein Dreieck, wobei die eine Seite auf dem Wasser liegt. Diese außergewöhnliche Architektur ermöglicht eine maximale Besonnung der einzelnen Wohneinheiten mit freiem Blick aufs Meer. Die großen Sonnenterrassen richten sich nach Süden zum Land hin. Hinter den Panoramafenstern befinden sich die weit hervorstehenden Wohnzimmer, die jeweils die doppelte Wohnungsbreite umfassen. Die Fassaden aus ungeschliffenen, rauen Aluminiumelementen sollen das Grau des niederländischen Wassers und Wetters widerspiegeln.

斯芬克斯住宅

20世纪90年代中期，在阿姆斯特丹附近霍伊湖南岸，努特林斯·雷代克建筑师事务所设计了8栋住宅楼，在同一时期，增加了5栋完全建在水中的住宅，它们赫然出现在水中，外形就像斯芬克斯一样令人激动。设计师们构想的建筑横断面就像一个三角形，因此一条边在水中，这种独特的结构保证了每套住宅单元都能最大程度地朝向开阔的海面。宽大的南向阳台朝着陆地方向，巨大的全景玻璃窗位置是出挑深远的起居室，其周长是公寓宽度的两倍，建筑外表的亚光金属铝板反射着海面和天空的颜色。

Wide spacious asphalt bridges lead from the shore to the impressive houses. Their form is reminiscent of giant sphinxes, after which they are named. The main difference is, however, that these sphinxes are not located in the boiling heat of the desert, but instead in a completely calm lake in the Netherlands.

Breite, großzügige und asphaltierte Stege führen vom Ufer zu den beeindruckenden Wohnhäusern. Ihre Form erinnert, wie der Name bereits sagt, an riesige Sphinxen. Nur das diese nicht in der flirrenden Hitze einer Wüste stehen, sondern auf einem spiegelglatten See in den Niederlande.

宽阔的柏油路面引桥连接湖岸和令人印象深刻的住宅，它们的形式让人联想起巨大的斯芬克斯。它们之间的区别在于，这些建筑不是在滚烫的沙漠之中，而是在荷兰平静的湖水之中。

奥斯卡·尼迈耶

Publishing House Mondadori

Oscar Niemeyer proved back in 1968 with his design of the publishing house Mondadori in Italy that great buildings do not require use of the right angle. The surprisingly different arches of this Milanese publishing house do create a cube but are truly impressive in the way they dissolve the rigidity of this form. The columns merge seamlessly through their reflection with the calm water that gently envelops their base. The irregular distances of the arches create an elegant tension that is transferred to the water surface and that incorporate the free space in front of the building in the outline without using it directly.

Verlagsgebäude Mondadori

Mit dem Verlagsgebäude Mondadori stellte Oscar Niemeyer schon 1968 evident unter Beweis, dass großartige Bauwerke nicht auf den rechten Winkel angewiesen sind. Die überraschend anders wirkenden Bögen des Mailänder Verlagshauses bilden zwar insgesamt einen Kubus, beeindrucken jedoch gerade dadurch, dass sie die Starre dieser Form gekonnt auflösen. Diese Säulen gehen mit der ruhigen Wasserfläche, die ihre Sockel umspielt, über ihr Spiegelbild eine unlösbare Verbindung ein. Die uneinheitlichen Abstände der Bögen erzeugen eine elegante Spannung, die sich auf die Oberfläche des Wassers überträgt und den freien Raum vor dem Gebäude in den Entwurf einbindet, ohne ihn physisch zu nutzen.

蒙达多里出版大厦

1968 年，奥斯卡·尼迈耶通过他设计的蒙达多里出版社大厦，证明了伟大的建筑不必依赖于直角。米拉内塞出版社令人诧异的不尽相同的拱门构成了一个“方盒子”，但真正令人印象深刻的，是这些拱门消除了（方盒子）这一形式的平庸单调和缺少变化。拱门与静水中的倒影非常自然的构成的一个整体，围绕基地的水面波澜不惊，拱门彼此之间不规则的间距形成一种优雅的张力通过倒影反映在水面上，这种效果使倒影前方的虚无空间变得具体化。

With his architectural language of curved forms and the dramatic display of water in front of the building, Oscar Niemeyer dissolves the contrast to his favorite building material. The same approach is visible in the way he opposes the size of the publishing house and the seriousness of the workload by adding softly flowing, communicative elements.

Oscar Niemeyer löst mit den geschwungenen Formen und der Inszenierung der Wasseranlage vor dem Haus den Kontrast zu seinem bevorzugten Baustoff geschickt auf. Genau wie er der Größe des Verlagshauses und der Ernsthaftigkeit der hier herrschenden Arbeitswelt etwas Fließendes und Verbindendes entgegensetzt.

通过“弧线式”建筑语汇和水面戏剧化的表现，奥斯卡·尼迈耶化解了他所钟爱的建筑材料之间的对立。他反对增加出版社大楼的设计规模，坚持增加那些具有流动性、可交流的设计元素（尽管这带来工作量），很明显的他以同样的方式处理上述两者之间的矛盾。

蒂亚戈·奥利韦拉

Hotel Estalagem da Ponta do Sol

The hotel *Estalagem da Ponta do Sol* on the southern coast of Madeira, Portugal, does not need any artificial distractions. The striking location on a steep craggy cliff over the Atlantic coast, the historical legacy of the old, renovated manor house and the reduced design speak a very unique language. White is the dominant color, as it does justice to the functional architecture of the new elements, the grain of the furniture and the floor and the earth tones of the old walls. Most importantly, however, the bright white high above the sea contrasts classically with the blue color of the water.

Hotel Estalagem da Ponta do Sol

Das Hotel *Estalagem da Ponta do Sol* an der Südküste Madeiras braucht keine künstlichen Augenschmeichler. Die Aufsehen erregende Lage auf einem steil abfallenden Felsvorsprung über der Atlantikküste, das historische Erbe des alten, renovierten Herrenhauses und das reduzierte Design sprechen eine ganz eigene Sprache. Weiß ist die beherrschende Farbe. Sie lässt die funktionale Architektur der neuen Elemente ebenso gut zur Geltung kommen wie die Maserungen des Mobiliars und der Böden und die erdigen Töne der alten Mauern. Vor allem aber bildet das strahlende Weiß hoch über dem Meer einen wunderbaren klassischen Kontrast zum Blau des Wassers.

索尔角酒店

位于葡萄牙所属马德拉群岛的索尔角酒店不需任何人工的粉饰，它建于陡峭的绝壁之上，位置极其壮观，俯瞰大西洋海岸古老的历史遗迹、翻新的庄园，简约的设计，手法独特。白色是主导色，它与新增的功能建筑、家具和地面的纹理、土色的古老墙面非常相配。更主要的是，海面上的亮白色和水池中的蓝色形成了绝妙的对比。

The severe lines and the angles of the outline have been successfully achieved through defining shadows and well-conceived mirror effects. The water in the pool is thus not only an airy bridge to the Atlantic Ocean, but also a projection surface for modern design.

Der Entwurf überzeugt nicht nur durch seine stringente Linienführung sondern auch durch die wirkungsvolle Schattengebung und die wohl durchdachten Spiegeleffekte. So bildet das Wasser des Pools nicht nur eine luftige Brücke zum Atlantik, sondern wird zur Projektionsfläche modernen Designs.

明确的阴影和倒影效果使得建筑严谨的轮廓和鲜明的形象得以实现。池水不仅是连接大西洋的虚幻之桥，也是现代设计的一种投影。

PURPUR 建筑师事务所和维托·阿孔奇

Aiola Island Café

Just like a glass sculpture, this artificial island lies in the middle of the Mur River in Graz (southeastern Austria). International star designer Vito Acconci inspired by an idea by Robert Punkenhofer dreamt up a weblike, 154 feet (47 Meter) long steel structure. The curved, wound form is reminiscent of a half-opened seashell. Through this exceptional design, two decisive zones are created which flow into one another seamlessly. One is the *Dome,* which is where the café is located, covered by a daring glass-steel construction. The *Bowl,* however, is the recreational area of the island—an arena for various activities. A playground unites both elements. The different areas were deliberately not separated from each other. Instead, they are in harmony with their flowing and changing surroundings, the Mur River.

Aiola Island Café

Wie eine gläserne Skulptur liegt diese künstliche Insel in der Mur in Graz und wird vom Wasser des Flusses umspielt. Der internationale Design-Star Vito Acconci hat nach einer Idee von Robert Punkenhofer eine netzartige, 47 Meter lange Stahlkonstruktion erdacht. Die geschwungene, gedrehte Form erinnert an eine halb geöffnete Muschel. Durch diesen außergewöhnlichen Entwurf entstehen zwei markante Zonen, die fließend ineinander übergehen. Das eine ist der „Dome“, in dem sich ein Café befindet und der von einer kühnen Glas-Stahl-Konstruktion überdacht wird. Die „Bowl“ dagegen ist der Freibereich der Insel: eine Arena für verschiedene Veranstaltungen. Im Drehpunkt dieser beiden Elemente befindet sich ein Spielplatz. Die verschiedenen Bereiche sind bewusst nicht voneinander getrennt worden. Sie korrespondieren vielmehr mit der fließenden und veränderlichen Umgebung, der Mur.

艾奥拉咖啡岛

这座人工岛坐落在格拉茨（奥地利南部）穆尔河的中间，国际明星设计师维托·阿孔奇创作的 154 英尺（147 米）长的钢结构就像一座玻璃雕塑，其创意来自罗伯特·蓬肯霍费尔所构想的一个“网状”物体。曲线形的敞口形式使人联想到半开的贝壳，通过这一独特的设计，小岛完美地形成了两个彼此相连的区域。一个是“盖”，内部为咖啡室，它覆盖着一个造型大胆的由玻璃和钢组成的顶盖结构；一个是“碗”，属于休闲区域——可以作为多种活动的场所。一个平台将两部分连成一体，两个外表不同的区域通过设计浑然一体、彼此难分。它们与周边不断变化的场景——流动的穆尔河非常和谐。

As in a spiral, the entrance faces the bench, which, in turn, is screwed to the bar. The flowing transitions in the interior correspond to the natural water movements of the Mur River.

In einer Schraubenbewegung verdreht sich der Eingang in die Sitzbank, die Bank zur Bar. So korrespondieren die fließenden Übergänge im Inneren mit den natürlichen Wasserbewegungen der Mur.

通过螺旋状的构造设计，使得朝着河岸方向的入口转过来后对着咖啡区，流畅的过渡和穆尔河水流的运动彼此呼应。

伦德尔、帕尔默和特里顿建筑师事务所

Thames Flood Barrier

The construction in the London suburb Woolwich might seem rather futurist; it serves, however, a very profane purpose. The 1715 feet (523 meter) wide *Thames Flood Barrier* protects London from storm flooding. Inopportune winds could push massive water masses up the river were it not for the anchored machine towers and one-sided arched gates. To bring the gates up takes less than 30 minutes. The middle gates have a width of 169 feet (60 meters), a height of 34 feet (10.5 meters) and a weight of 1,500 tons. Ship traffic normally passes through these four gates. If all 10 gates of the largest moving storm flood barrier in the world are closed, then the Thames' wide mouth is completely closed. The barrier is re-opened by sinking the gates into the riverbed, which is done through a rotative element seal with the arched side facing downwards.

Thames Flood Barrier

Was in der Nähe des Londoner Vororts Woolwich futuristisch anmutet, erfüllt einen ganz profanen Zweck: Die 523 Meter breite *Thames Flood Barrier* schützt London vor einer Sturmflut. Ungünstige Winde könnten gewaltige Wassermassen flussaufwärts drücken, würden nicht die zwischen den Maschinentürmen verankerten und einseitig gewölbten Tore hochgedreht. Das geschieht in weniger als 30 Minuten. Die mittleren Tore haben jeweils eine Breite von 60 Metern, eine Höhe von 10,5 Metern und ein Gewicht von 1.500 Tonnen. Diese vier Tore passiert normalerweise der Schiffsverkehr. Werden alle zehn Tore der größten beweglichen Sturmflutbarriere der Welt geschlossen, ist die breite Mündung der Themse komplett gesperrt. Die Sperre wird wieder geöffnet, indem die Tore durch einen Drehsegmentverschluss mit der gewölbten Seite nach unten in das Flussbett versenkt werden.

泰晤士河防洪闸

在伦敦郊区伍利奇的这座构筑物看起来非常具有未来感，但实际上它有着非常实际的功能。这座 1715 英尺（523 米）宽的泰晤士河防洪闸可以保护伦敦免受暴雨洪水之灾。季风会推动大量的水聚集在河道之中，但只需不到 30 分钟的时间锚机塔和单向拱门就会升起并将它们拦下。中门宽 169 英尺（60 米）、高 34 英尺（10.5 米）、重 1500 吨，通常情况下有 4 个闸口开启供通行船只，如果世界上最大的 10 个移动式防洪闸全部关闭，宽阔的泰晤士河口将被彻底封锁。闸口打开时，通过回转机械装置，拱形的闸门将面朝下沉入河床。

8

Like large shark fins, the stainless steel enveloped gates tower out of the Thames' waters.

Wie riesige Haifischflossen ragen die mit Edelstahl ummantelten Maschinenhäuser aus dem Wasser der Themse.

浮现在河面上的不锈钢闸门塔就像巨大的鲨鱼鳍。

Water has been tamed here for centuries.

Hier wird seit Jahrzehnten eindrucksvoll Wasser gebändigt.

几个世纪以来，河水在这里被防洪闸所驯服。

米罗·里韦拉建筑师事务所

Lake Austin Boat Dock

Juan Miró and Miguel Rivera have designed the perfect boat dock for the forested shoreline of Lake Austin in Texas. Built on a steel frame, the *Lake Austin Boat Dock* offers facilities for two boats. The architects designed the driveway for the boats by positioning them on the two narrow sides, so that the boats face each other in the bow, instead of being side by side. The boat house's design not only appears light and dynamic, but can also be used as a bathing jetty. The front consists of parallel place steel pipes, which offers a degree of privacy and ample amount of fresh air and sunlight. This creates a superficial structure, which seems to move in synchronicity with the water surface.

Lake Austin Boat Dock

Juan Miró und Miguel Rivera haben mit dem *Lake Austin Boat Dock* einen Bootsanleger entworfen, der sich der bewaldeten Küstenlinie des Lake Austin in Texas perfekt anpasst. Der aus einem Stahlrahmen konstruierte Bootsanleger beinhaltet Slip-Anlagen für zwei Boote. Die Architekten haben die Einfahrten für die Boote an den beiden Schmalseiten positioniert, so dass die Boote einander mit dem Bug gegenüberliegen und nicht wie gewohnt Seite an Seite. Das Design des Bootshauses, das zugleich als Badesteg genutzt werden kann, wirkt leicht und dynamisch. Die Front besteht aus waagerecht angebrachten Stahlrohren, die einerseits vor unerwünschten Blicken schützen und andererseits Luft und Licht hereinlassen. Dadurch entsteht eine Oberflächenstruktur, die scheinbar gemeinsam mit dem Wasserspiegel in Bewegung versetzt wird.

奥斯汀湖游船码头

位于得克萨斯州奥斯汀湖的森林海岸线游船码头非常完美，它由胡安·米罗和米格尔·里韦拉设计。游船码头建在一个钢架结构之上，可供两艘船停靠。建筑师为游船设计的停靠方式是把船布置在两条窄边上，这样两船的船头彼此相对而不是船体彼此相邻。船坞的设计不仅体现出轻盈和动感，在功能上还可以作为游泳码头。它的正面是一面平行的钢管挡墙，可以保证一定程度的隐私，让足够的新鲜空气和光线透过。倒映在湖面之上的船坞，好似随着水面一起上下波动。

The upper deck is protected by a large sun sail, which is stretched out between steel posts.

Das Oberdeck wird von einem großzügigen Sonnensegel beschattet, das zwischen Stahlmasten gespannt wird.

船坞顶部的平台上，在钢柱之间拉伸而成的遮阳棚面积很大。

The sun sail is not only stable, but also resistant to wind and water.

Das Sonnensegel ist nicht nur stabil, es widersteht auch Wind und Wasser.

遮阳棚不仅坚固，而且能遮风挡雨。

RMP 斯蒂芬·伦岑景观建筑事务所

T-Mobile Stadt

Landscape architect Stephan Lenzen designed an exceptional concept from the headquarters of this business in Bonn, Germany. A total of five office complexes are grouped around a communal square and forum. The main theme of the inner courtyards is water surface as a mirror of the façades. Circular islands are designed as green landscapes, and the larger islands can be reached through stepping stones and used as a rest and relaxation area. Each inner courtyard features a large wooden deck as a recreational area for the employees. Even the urban-defined area is inviting as a place of rest and communication. Placed between the natural stone slabs adding an artistic touch, 27 water nozzles spray water into the air.

T-Mobile Stadt

Für den Hauptsitz des Unternehmens in Bonn entwarf der Landschaftsarchitekt Stephan Lenzen ein außergewöhnliches Konzept. Insgesamt fünf Büroquartiere gruppieren sich um einen gemeinschaftlichen Platz und das Forum. Das Grundthema der Innenhöfe bilden Wasserflächen als Spiegel der Fassaden. Kreisförmige Inseln sind als begrünte Landschaften angelegt. Die größeren Inseln können über Trittsteine als Verweil- und Ruhebereich genutzt werden. In jedem Innenhof befindet sich ein großes Holzdeck als Aufenthaltsbereich für die Mitarbeiter. Auch der städtisch geprägte Platz lädt zum Verweilen und zur Kommunikation ein. Ein künstlerisches Element sind die 27 Wasserdüsen, die zwischen den Natursteinplatten Wasser in die Luft sprühen.

T–Mobil城

景观建筑师斯蒂芬·伦岑为T–Mobil公司在德国波恩市的总部所做的设计概念独特。五栋办公建筑围绕一个公共广场和论坛进行布局组织，内部庭院以水为设计主题——它像镜子般反射着建筑外墙。圆形的小岛上种有绿色植物，踏脚石所连接的大岛则作为休息和放松的场所。每个庭院的内部，有大面积的木平台供雇员们活动。即便是属于城市范围的区域也被赋予休闲和交流功能。在天然石材铺装地面上的27个喷泉向空中喷射着水花，无形中增加了一些浪漫气息。

桑努+埃娃·史弗达萨尼

Soneva Fushi

Sonu and Eva Shivdasani, founders of the *Six Senses Resorts & Spas,* have enriched the Maldives with an exceptional resort, the *Soneva Fushi.* With a length of 4,593 feet (1400 meters) and a width of 1,312 feet (400 meters), Kunfunadhoo in North-Baa-Atoll is one of the larger islands of the Maldives. With its 65 unusual villas and suites, it is a direct invitation to take time to relax, surrounded by the ocean. Soneva Fushi means 'contemplating nature.' The sophisticated residences are in lush gardens. Many have a private pool, which encircles the building. Walking barefoot on the sand and watching a movie out in the open at *Cinema Paradiso*—all this is done with the soft murmur of water in the background.

Soneva Fushi

Sonu und Eva Shivdasani, Gründer der *Six Senses Resorts & Spas,* haben mit *Soneva Fushi* die Malediven um ein Resort besonderer Art bereichert. Mit 1.400 Metern Länge und einer Breite von 400 Metern ist Kunfunadhoo im Nord-Baa-Atoll eine der größeren Inseln der Malediven und lädt mit 65 ungewöhnlichen Villen oder Suiten zu einer Auszeit inmitten des Meeres ein. *Soneva Fushi,* das bedeutet im doppelten Sinn Besinnung auf die Natur. Die stilvollen Unterkünfte liegen in üppigen Gärten. Viele haben einen privaten Pool, der sich sanft an die Gebäude anschmiegt. Barfuss über sandige Wege laufen und im *Cinema Paradiso* unter freiem Himmel einen Film schauen – alles geschieht vor dem Hintergrund leisen Meeresrauschens.

撒内瓦·弗士度假村

桑努和埃娃·史弗达萨尼是第六感度假和水疗集团的创始人，他们以别具特色的撒内瓦·弗士度假村，让马尔代夫变得更加丰富多彩。长 1312 英尺（1400 米）、宽 4593 英尺（400 米）的库凡那达荷是马尔代夫北巴阿环礁最大的岛屿之一。65 套风格独特的别墅套房位于大海之中，让人流连忘返。撒内瓦·弗士意为“绝对的自然”，精巧的建筑掩映在绿树之间，其中有很多环绕着房屋的专属泳池。无论是赤脚漫步在沙滩上，还是在户外欣赏电影，海水的喃喃细语一直相伴左右。

Guests have direct access to the sea from almost each residence. Discovering the underwater world while diving and snorkeling, riding the waves, going water skiing, windsurfing or simply swimming–the ocean is never more than 80 feet (25 meters) away from the vacation paradise.

Von fast jeder Unterkunft haben die Gäste einen direkten Zugang zum Meer. Beim Tauchen und Schnorcheln die Unterwasserwelt entdecken, Wellenreiten, Wasserski fahren, Windsurfen oder einfach nur schwimmen: Das Meer ist nie mehr als 25 Meter vom Urlaubsrefugium entfernt.

几乎每栋建筑的客人都可以直接下水，潜入水中探索水底世界、冲浪、滑水、风帆或只是游个泳——在这座度假村，海水永远在80英尺（25米）之内。

Soneva Fushi boasts many spectacular sites, both in and out of the water. One example specifically designed for the surroundings, is a whirlpool directly immersed in the sea. Not only ideal for a swim, it offers a breathtaking view—it is an exceptional lounge in the water.

Soneva Fushi verfügt über außergewöhnliche Plätze am und im Meer. Mit einem der Umgebung angepassten Design findet man auch einen Whirl-Pool im Meer. Hier wird nicht nur gebadet, hier sitzt man und genießt den Blick – die Lounge am Wasser.

撒内瓦・弗士度假村拥有的许多特别之处都和水有关。上面是个特别的、结合所在环境而设计的例子，它仿佛是一个从海面直接升起的涡旋浴池，在其中你不仅可以戏水放松，它还有无敌的景色——就像一个海上会客厅！

lot 工作室

Hofgut Hafnerleiten Guest Huts

The *Hofgut Hafnerleiten*, embedded in the lower Bavarian sloping landscape, offers its guests, in addition to the exclusive cooking school, six very special sleeping facilities for nature lovers. All guest houses are identical in size, proportion, roof shape and façade cladding, only their surroundings vary. The inner rooms were designed individually in order to highlight the special character of each hut. Each of the six guest houses are dedicated to a special theme. The Water House, for example, is on stilts, in the middle of a small swimming lake in the tranquil territory of the complex. Wooden piers connect the house with the surrounding landscape. Large-sized windows let the fantastic water landscape seemingly flow into the interior.

Gästehütten Hofgut Hafnerleiten

Das *Hofgut Hafnerleiten*, eingebettet in die niederbayrische Hügellandschaft, bietet seinen Gästen, neben der exklusiven Kochschule, sechs besondere Übernachtungsmöglichkeiten für Naturliebhaber. Alle Gästehäuser sind in Größe, Proportion, Dachform und Fassadenbekleidung identisch, sie unterscheiden sich lediglich durch ihre Umgebung. Die Innenräume wurden individuell gestaltet um die Besonderheit einer jeden Hütte herauszuarbeiten. Je nach Lage sind die sechs Gästehäuser einem bestimmten Thema zugeordnet. So steht das Wasserhaus zum Beispiel auf Stelzen, mitten in einem kleinen Badesee im sanften Gelände der Anlage. Holzstege verbinden das Haus mit der umgebenden Landschaft. Großformatige Fenster lassen die phantastische Wasserlandschaft scheinbar bis in den Innenraum fließen.

霍夫特・哈芬尔滕客舍

霍夫特・哈芬尔滕项目坐落在下巴伐利亚的丘陵景观之中，除了一所专门的烹饪学校外，还有6栋特别的建筑供户外爱好者居住。所有房屋的尺寸、面积、屋顶，还有立面形式都一样，只是所处位置环境不同。房间的内部有不同的设计，各自突出的特色。每栋房屋有自己的主题，例如“水屋”，所在位置很安静，建在一个可以游泳的小湖中，有木制栈桥连接湖岸，巨大的玻璃窗让美丽的湖景一览无遗。

TANGRAM 建筑师事务所

Water Houses

Nine blocks, with two houses each, is the structure of the residential complex on the Dutch coast. In order to avoid disturbing the biosphere water, the houses seal in the water surface with only one third of their surface structure. The green roofs offer additional space for birds and other animals, while the thatched construction offers a high degree of flexibility when creating the interior. Thanks to the character of the choice of material, especially wood and concrete, and the efficient use of natural daylight, these houses are exceptionally energy-saving. The southern side with the sun-room opens almost completely to the exterior, while the northern side is kept almost completely closed. The exterior of each house is structured in a terrace on the level of the living area, which gives a wonderful view out onto the Dutch water channels and onto a docking pier on the water. These decks, separated from the neighboring properties with fences, offer the necessary sense of privacy.

Wasserhäuser

Neun Blöcke mit jeweils zwei Häusern – so gliedert sich diese Wohnanlage an der holländischen Küste. Um den Lebensraum Wasser nicht zu stören, versiegeln die Häuser nur mit einem Drittel ihrer Grundfläche die Wasserfläche. Die begrünten Dächer bieten zudem Raum für Vögel und andere Tiere. Die Fachwerk-Konstruktion bietet eine hohe Flexibilität bei der Gestaltung der Innenräume. Die Häuser sind auf Grund der Materialwahl, vor allem Holz und Beton, und der gewinnbringenden Nutzung des natürlichen Tageslichts ausgesprochen energiesparend geplant. Die Südseite mit ihrem Sonnenraum öffnet sich fast vollständig nach Außen, während die Nordseite fast komplett geschlossen gehalten wurde. Der Außenbereich jedes Hauses gliedert sich in eine Terrasse auf der Wohnebene, die einen wunderbaren Blick auf die Wasserkanäle erlaubt und einen Anlegesteg. Diese Decks werden durch Reetzäune von den benachbarten Grundstücken getrennt und bieten so die nötige Privatsphäre.

水住宅

这一位于荷兰海滨的居住项目有9个组群，每组包括2栋房屋。为了避免干扰水的生态模式，位于水平面上方、结构外露的建筑部分中，仅有三分之一作为房间使用。环保型屋面给鸟类和其他动物提供了额外的活动空间，桁架式结构也为室内空间的设计提供了足够的灵活性。得益于对建筑材料的选择，特别是木材和混凝土，还有对日光的高效应用，建筑的能耗非常低。南向的日光房几乎完全对外开敞，北面则几乎是封闭的。起居区域所在的建筑部分，其外表采用了柱廊的形式，这种形式也用于水面上的木平台，从海上看非常美观。栅栏将木平台和住户的房屋相隔离，提供了必要的私密感。

蒂格内斯图恩·范德昆斯坦

Torpedo Hall Apartments

The former Torpedo Hall in Holmen near Copenhagen, Denmark, has been transformed into an exceptional residential complex. With countless channels, the maritime atmosphere and the historical warehouses out of red brick, the former docking facilities of the Royal Danish Navy offer an unbelievable background for 67 apartments and lofts. The architects wished to keep the hall's original construction during the renovation. Therefore, they inserted a unique concrete structure, which is based on the measurements of the old warehouse, and removed the roof to let additional light stream into the interior. An interior 'street' connects the apartments. In addition, a wooden stairway leads down to the lower floors and to the channel. Residents can dock here with their own boats.

Torpedo Hall Apartments

Die ehemalige Torpedohalle in Holmen bei Kopenhagen ist in eine außergewöhnliche Wohnanlage verwandelt worden. Mit den zahlreichen Kanälen, der maritimen Atmosphäre und den historischen Lagerhallen aus rotem Klinker bilden die ehemaligen Dockanlagen der Königlichen Marine eine besondere Kulisse für die 67 Apartments und Lofts. Die Architekten wollten beim Umbau die ursprüngliche Konstruktion der Halle beibehalten. Sie setzten eine eigene Betonstruktur, die sich an den Maßen der alten Halle orientiert, hinein und entfernten das Dach, um mehr Licht ins Innere zu bringen. Über eine innen liegende „Straße“ werden die Wohnungen erschlossen. Eine hölzerne Treppe führt zu den unteren Stockwerken und dem Kanal. Hier können die Bewohner mit ihren eigenen Booten andocken.

鱼雷厂公寓

位于丹麦哥本哈根附近何蒙恩霍尔门的前鱼雷厂，已被改造成独特的居住建筑。这里曾有不计其数的管道、海港式的氛围、久远的红色砖墙仓库，原有的皇家丹麦海军船坞码头令人难以相信地变成了 67 套公寓和 loft 的载体。建筑师希望在更新改造中能保留船坞的原始结构，因此，他们在对旧有仓库进行评估后加入了特殊的混凝土结构，改造了屋面，以引入更多日照进入内部空间。一条内部“街道”将这些公寓进行联系，此外，一座木踏板楼梯通向底层和过道，住户可以将自己的船直接停靠在这里。

Bridges, piers and stairs with glass balustrades run in and along the house and permit a variety of views, both inside the complex and outside at the surrounding water.

Brücken, Stege und Treppen mit verglasten Brüstungen laufen im Haus und am Haus entlang und erlauben so vielfältige Ein- und Ausblicke in die umgebende Wasserlandschaft.

装饰以玻璃栏板的天桥、通道和楼梯在建筑内部交错穿插，使得楼内和外部水面的景色产生丰富的变化。

All apartments have balconies or terraces which face the water.

Alle Wohnungen verfügen über Balkone oder Terrassen, die zum Wasser hin liegen.

所有住户都拥有面向大海的阳台或露台。

Water Villas

The city Almere has been evolving since 1975 and is located on the southern part of the Flevolandpolders, around 25 kilometers east of Amsterdam. On the border of the Zuidersees, a new building experiment with 48 studios has been developed, based on a design by UNStudio. In order to fulfill the owners' individual requirements, the architects used a basic foundation for each house, which can be adapted or modified to the desired form by adding on protruding 'boxes' positioned above or below the foundation. In this way, a flexible, modular system was created where different levels and volume bodies are interconnected. Terraces, balconies and open areas face towards water. Water even flows between the maze of villas and creates its own little channel system. This is what ultimately characterizes this design—all buildings are surrounded by water and the reflections in the smooth water surface create a very special atmosphere.

Wasservillen

Die Stadt Almere wird seit 1975 auf dem südlichen Teil des Flevolandpolders, ca. 25 km östlich von Amsterdam, gebaut. Am Rande des Zuidersees entstand ein Wohnexperiment mit 48 Wasservillen nach einem Entwurf von UNStudio. Um die individuellen Vorstellungen der Besitzer zu erfüllen, besteht jedes Haus aus einem Basisfundament, das durch vorstehende, aufgestellte oder untergebaute „Boxen" in der gewünschten Form verändert werden kann. So entstand ein flexibles, modulares System, bei dem verschiedene Ebenen und Volumenkörper verbunden werden können. Um die natürlichen Gegebenheiten mit einzubeziehen, entstanden Terrassen, Balkone und Freiflächen, die sich zur Wasserfläche hin öffnen. Auch zwischen den verschachtelten Villen fließt Wasser und schafft ein eigenes kleines Kanalsystem. Gerade dies zeichnet den Ort aus; alle Gebäude sind von Wasser umgeben und die Spiegelungen auf der glatten Wasseroberfläche schaffen eine ganz besondere Atmosphäre.

水边别墅

阿尔梅勒城位于弗莱福兰珀德斯南部、阿姆斯特丹以东约 25 公里处，自 1975 年以来，它一直在扩张。在祖德西斯的边缘地带，一处新的、包括有 48 栋工作室的实验性建筑落成了，项目由 UNStudio 设计。为了满足业主各自的需求，针对每栋房屋设计师都会先采用一种基本框架，然后通过在框架上方或下方增加凸出的“盒子”以适应所需。通过这一方式，不同标高和体积的模块彼此连接，形成了弹性的、模块化的系统。门廊、阳台和公共区都朝着水面的方向，在迷宫似的别墅区之中嵌套着水道，形成微型的水系。这是设计的根本特质——水环绕着所有的建筑物，水面倒影形成了别致的画面。

密斯·凡·德·罗

Barcelona Pavilion

The so-called *Barcelona Pavilion* is generally understood to be one of the most exceptional buildings of modern times which has inspired generations of architects. It was created in 1929 for the World Fair in the Spanish metropolis Barcelona. The revolutionary aspect of this building style is the dissolution of walls in their function as a statistical element. Supporting beams out of steel carry the protruding roof. Floors and wall surfaces do not surround the room; They merely function as room dividers. Two pools embedded in the ground are outside. The front seems to be a large mirror reflection, which conveys meditative peace. While strolling through the pavilion, visitors discover a second pool behind the room-high glass wall. Together with three marble walls, it functions as a platform for the bronze figure 'The Morning' by the artist Georg Kolbe, which is reflected in the waters.

Barcelona Pavillon

Der so genannte *Barcelona Pavillon* wird als eines der herausragendsten Gebäude der Moderne verstanden und hat bis heute Generationen von Architekten inspiriert. Entstanden ist er 1929 zur Weltausstellung in der spanischen Metropole. Das Revolutionäre an seiner Bauweise ist die Loslösung der Wände aus ihrer Funktion als statisches Element. Stahlstützen tragen das weit vorspringende Dach, Fußboden und Wandfläche umschließen den Raum nicht, sondern sind nur Raumteiler.
Im Außenbereich befinden sich zwei in den Boden eingelassene Wasserbecken. Das Vordere wirkt wie eine große Spiegelfläche, die meditative Ruhe verströmt. Beim Durchgang durch den Pavillon entdeckt der Besucher hinter einer raumhohen Glaswand ein zweites Becken, das durch drei Marmorwände begrenzt, als Plattform für die sich spiegelnde Bronzefigur „Der Morgen“ des Künstler Georg Kolbe dient.

巴塞罗那世博会德国馆

所谓的“巴塞罗那馆”被公认为最具代表性的现代主义建筑之一，直到今天它仍在激励一代又一代的建筑师。1929年在西班牙巴塞罗那举行的世界博览会期间建成。这栋建筑革命性的举措在于——墙体不再作为支撑要素，由钢柱支撑着出挑的屋面，地面、墙面的布置不再受制于房间自身，它们只是起着划分空间的作用。两个建在地面的室外水池中，前面的水池像一面镜子，散发着冥想般的宁静，观众穿过展馆后，会在一面通顶玻璃墙后看到第二个水池，还有三面大理石墙面，后者成了舞台背景，艺术家格奥尔·科尔比创作的青铜雕塑“清晨”静立在水池中。

马克斯·韦贝格

Kraftwerk Spa Autostadt

A visit to the Volkswagen *Autostadt* in Wolfsburg, Germany, is no longer something that only Volkswagen fans will enjoy. In fact, there is something for everybody's taste here at the *Autostadt.* One special highlight is the *Kraftwerk Spa* in the *Ritz-Carlton,* which promises various ways to achieve ultimate rest and relaxation. Visitors can enjoy all luxuries of a modern-day spa area—Jacuzzis, saunas and a fitness area. Yet the exterior of the complex is truly exceptional. Connected to a wooden stairway with a wooden deck lies an extended terrace with a pool, similar to an island in the port basin. The pool's water temperature is always 82.4°F (28°C), so swimming is possible in summer and winter. Moreover, the views out onto the channels and the historical power plant are stunning.

Kraftwerk Spa Autostadt

Ein Besuch der VW-Autostadt ist nicht mehr nur Fans der Wolfsburger Marke vorbehalten. Längst liefert die Autostadt für jeden das Passende. Ein besonderes Highlight ist das *Kraftwerk Spa* im *Ritz-Carlton,* das exklusive Erholung und Entspannung verspricht. Hier findet der Besucher alle Annehmlichkeiten eines modernen Wellnessbereichs: Jacuzzis, Saunen und ein Fitnesszentrum. Das Besondere sind aber die Außenanlagen. Eine lang gestreckte Terrasse mit einem Schwimmbecken liegt wie eine Insel im Hafenbecken. Eine hölzerne Treppe stellt die Verbindung zum Holzdeck dar. Bei 28°C ist das Baden im Sommer wie im Winter möglich. Und der Blick auf die Kanäle und vor allem auf das angrenzende historische Kraftwerk ist einmalig.

汽车城卡夫特维尔克 Spa

不再只有大众车迷才对参观沃尔夫斯堡大众汽车城感兴趣。实际上，汽车城的有些东西适合所有人的口味。一个特别的亮点就是丽思－卡尔顿酒店的卡夫特维尔克 Spa，它提供了很多种令人彻底放松和休息的方式。参观者将尽享当今 Spa 的奢华——“极可意”按摩浴缸（Jacuzzis）、桑拿浴和康复健身。建筑的外部很特别，一座铺着木甲板的扶梯通向一个长长的、像港湾中小岛一样的水池，池水的温度始终保持在 28℃，无论夏季还是冬季，你都可以畅游其中。还有就是——运河对面古老的电力厂房令人难忘。

Large loungers on the sundeck are a direct invitation to lie down and relax.

Großzügige Liegen auf dem Sonnendeck laden zum Entspannen ein.

日光平台上宽大的躺椅，邀请你躺下放松。

Two seemingly antique bathing tubs hide modern whirlpools.

In zwei historisch anmutenden Badezubern befinden sich moderne Whirlpools.

实际上这两个看起来很老套的浴桶是最先进的涡轮按摩水疗设备。

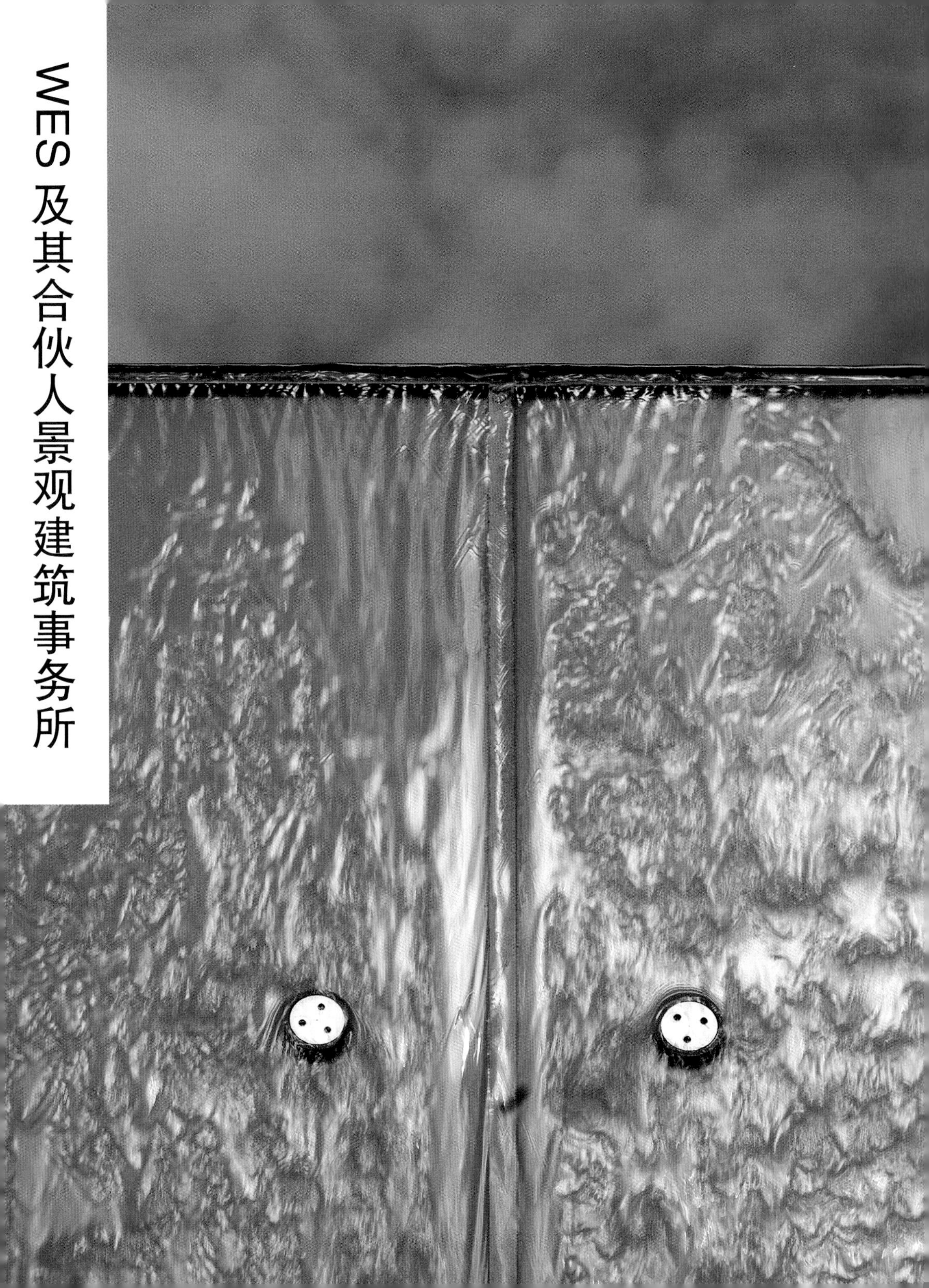

WES 及其合伙人景观建筑事务所

Penthouse Garden

A truly exceptional garden—such was the client's request. By extending the loft-like living area out into the open, an open-air room was created, with the heavens and the stars as a ceiling. Indeed, it could even be seen as a carefully staged show, as a type of theatrical composition. While the approach is minimalist, the Penthouse Garden contains all elements of a living area. A polished white marble floor stretches from the interior to the exterior, straight up to the green lawn carpet. An elegant, high-gloss varnished closet offers storage room, while simultaneously hiding the water technology. Thirteen feet (4 meter) high semitransparent curtains shimmer in their silver glory in the wind. A further focal point is the paneled wall pane out of satined glass, which is covered by a thin water film. Three multi-trunked shadberries, whose bizarre trunks resemble sculptures, project graphic shadows on the floor and onto the water wall.

Penthousegarten

Ein besonderer Garten sollte es werden, so der Wunsch des Bauherrn. Durch die Erweiterung des loftartigen Wohnraums ins Freie ist ein Raum entstanden, mit dem Himmel und den Sternen als Decke. Eine Art Bühnenkomposition, eine vorsichtige theatralische Inszenierung. Im Ansatz minimalistisch, enthält der Penthousegarten alle Elemente eines „Wohnraumes". Ein weißer, polierter Marmorfußboden zieht sich von Innen nach Außen bis hin zu dem echten grünen Rasenteppich. Hinzugefügt wurde ein eleganter hochglanzlackierter Schrank, der Wassertechnik und Stauraum verbirgt und vier Meter hohe Vorhänge. Ein weiterer Blickfang ist eine mit satiniertem Glas verkleidete Wandscheibe, die von einem dünnen Wasserfilm überzogen wird. Drei mehrstämmige Felsenbirnen, deren bizarre Stämme wie skulpturale Grafiken wirken, projizieren grafische Schatten auf den Boden und die Wasserwand.

阁楼花园

每个客户都想有个真正独特的花园。在这个项目中，设计师将阁楼式起居室的范围延伸至户外，形成了一个以天空和星星为天花的室外房间。实际上，你可以把它视为一种有着戏剧化成分的精致舞台秀。虽然设计师采用了极简主义的手法，但这个阁楼花园具备了起居室应有的所有要素。光亮的大理石地板从室内一直延伸到室外，绿色草坪如同地毯，典雅、闪亮的玻璃夹层涂有清漆形成储藏空间，隐藏着流水设备。13 英尺（4 米）高的半透明窗帘在微风中闪着银光。最为特别的地方是镶套着彩釉玻璃的玻璃墙壁，它的表面有一层薄薄的水幕在流淌。三丛外形奇特的唐棣树枝干如同雕塑，阳光下地面和水墙上树影斑驳。

Gurgling water and the alternating tree shadows create a calming atmosphere.

Plätscherndes Wasser und die alternierenden Baumschatten erzeugen eine beruhigende Atmosphäre.

潺潺的流水、交错的树影，共同营造出宁静温馨的氛围。

Semi-transparent and shiny silver curtains frame the Penthouse Garden.

Semitransparente und silbrig schimmernde Vorhänge rahmen den Penthousegarten.

半透明银色幕帘环绕着庭院外墙。

弗里多林·韦尔特

St. Arbogast Water Pavilion

In order to further integrate art in St. Arbogast, Austria, a competition was held dedicated to water. This resulted in a water pavilion out of concrete, designed by Fridolin Welte. Christian Lenz was responsible for the architectural execution of the outline. A 16 by 16 feet (5 by 5 meter) large concrete cube with 820 precisely drilled openings stands as a perforated monolith on the lawn, which comprises a rectangular water basin and a drinking area. The uncluttered forms and the peaceful calm of the water turn the pavilion into a place of peace and meditation. The roof sports the same regular openings, through which light, and also rain, heat, coldness and wind can pass. Inner and outer areas are closely intertwined and an exciting tension develops between open and closed and natural and sheltered elements.

Wasserhaus St. Arbogast

Um Kunst stärker in St. Arbogast zu integrieren, wurde ein Wettbewerb zum Thema Wasser ausgeschrieben. Der Ort erhielt dadurch einen Wasserpavillon aus Beton von Fridolin Welte. Mit der architektonischen Umsetzung des Entwurfs wurde Christian Lenz beauftragt. Der fünf mal fünf Meter große Betonkubus mit seinen 820 präzise gebohrten Öffnungen steht als perforierter Monolith auf der Wiese und umschließt ein rechteckiges Wasserbecken und eine Trinkstele. Die schlichten Formen und die beruhigende Stille des Wassers machen den Pavillon zu einem Raum der Ruhe und der Meditation. Im Dach setzen sich die gleichmäßigen Öffnungen fort und lassen nicht nur Licht, sondern auch Regen, Wärme, Kälte und Wind durch. So entsteht eine enge Verknüpfung zwischen Innen- und Außenraum und spannungsvolle Gegensätze zwischen Geschlossenheit und Öffnung, Schutzfunktion und Ausgesetztheit.

圣阿博加斯特水阁

在奥地利的圣阿博加斯特，为了进一步全面促进艺术的发展，当地举办了专以水为主题的设计竞赛。最终入选的是混凝土结构的"水阁"，该项目由弗里多林·韦尔特设计，克里斯蒂安·伦茨则负责设计的具体执行。混凝土立方体的尺寸为16英尺乘16英尺（5米乘5米），"水阁"就像旷野上布满孔洞的巨石，表面有820个整齐异常的钻孔。建筑内部有一长方形水池和一个饮水区。简洁的形式、祥和平静的池水，室内充溢着冥想般的氛围。屋顶和墙面一样，留有相同的孔洞，光、雨水、热气、冷气和风可以穿透建筑。室内和室外区域联系紧密、相互渗透，人在这种开阔与封闭、大自然和庇护所之间的冲突之中感受，兴奋之情悄然而生。

维尔克－萨利纳斯建筑师事务所

Badeschiff

The *Badeschiff* in Berlin, Germany, should be seen in the tradition of historical swimming pools on a boat, built at the turn of the 20th century. The pool consists of a restored and modified barge, the middle element of a push tow used in inland navigation shipping for transport. The barge was given the technology needed to transform it into a swimming pool, including water revolution, heating and water purification. The interior space was insulated and covered in blue swimming pool foil. Solidly anchored in the Spree River, this pool can be reached via an expansive bridge, the so-called Spree Bridge. A special membrane construction was used for the roof of the bathing boat, as to enable use during winter. In addition to the original pools, the facilities include saunas, a resting area and a bar.

Badeschiff

Das *Badeschiff* in Berlin reiht sich in die Tradition historischer Flussschwimmbäder der Jahrhundertwende ein. Das Becken besteht aus einem umgebauten Schubleichter, dem Mittelteil eines in der Flussschifffahrt zu Frachtzwecken genutzten Schubverbandes. Der Leichter wurde mit der für den Betrieb eines Schwimmbades erforderlichen Technik zur Umwälzung, Beheizung und Wasserreinigung ausgestattet und im Innenraum abgedichtet und mit blauer Schwimmbadfolie ausgeschlagen. In der Spree fest verankert, kann das Bad über eine ausgedehnte Steganlage, die so genannte Spreebrücke, erschlossen werden. Für eine Nutzung im Winter wurde das Badeschiff mit einer speziellen Membrankonstruktion überdacht und bietet nun neben dem eigentlichen Becken, Saunen, einen Ruhebereich und eine Bar an.

浴船

在德国柏林的浴船可被视为传统中的船上游泳池，它最早建于20世纪初期。泳池由一条经修复和改良的驳船中间部分构成，驳船之前被用于内陆航运。通过包括水循环、水加热和水过滤设施等必要的技术改造，驳船变成了游泳池。它的内部有隔热保温层，覆盖以蓝色泳池专用衬垫。驳船锚固在施普雷河中，一座宽阔的桥连接着泳池和河岸，名为“施普雷桥”。为了泳池在冬季也能使用，一种特殊的薄膜构造被用于顶盖。除泳池外，还有桑拿、休息区和酒吧等其他设施。

FEUER
WEHR

The *Badeschiff*'s roof consists of three light horizontal membrane constructions.

Drei längs gerichtete, leichte Membrankonstruktionen überspannen das *Badeschiff*.

浴船的顶盖是三个轻盈的跨越式膜结构。

Large-scale, transparent surfaces give a great view out onto the Spree River in the wintertime.

Großflächige, transparente Flächen ermöglichen den Ausblick auf die winterliche Spree.

在冬季，通过大面积的透明外表可以尽瞰施普雷河面的景色。

彼得·卒姆托

Therme Vals

The building of the *Therme Vals* seems almost sacred. Swiss architect Peter Zumthor designed an innovative building, whose quality emphasizes the relaxing powers of these healing Alpine springs. The building beautifully highlights the typical elements of the Alps, such as mountains, stones, water and nature. A total of 60,000 stone slabs from the locally occurring Vals quartzite were used during construction. Stacked on top of each other and bestowed with light slits, the walls evoke those of a cave. The thermal waters, originating from a hot spring in the mountain, are inserted in the austere, man-made cliff formation. Voices, steps, drops and the sound of water thundering echo off of the high walls. Small grottos are grouped around the central pool and create unique and beautiful worlds of their own.

Therme Vals

Fast sakral mutet das Gebäude der *Therme Vals* an. Der Schweizer Architekt Peter Zumthor hat ein innovatives Bauwerk entworfen, dessen Qualität den Erholungswert der alpinen Heilquelle zusätzlich unterstreicht. Das Bauwerk setzt auf besondere Art und Weise die typischen Elemente Berge, Steine, Wasser und Natur in Szene. Das Gebäude selbst besteht aus insgesamt 60.000 Steinplatten des örtlich vorkommenden Valser Quarzits. Aufeinander geschichtet und mit Lichtschlitzen versehen, wirken die Wände fast höhlenartig. Das Thermalwasser, das aus einer warmen Quelle dem Berg entspringt, fügt sich in die karg anmutende, menschlich geformte Felsformation ein. Stimmen, Schritte, Tropfen, Wasserrauschen hallen von den hohen Wänden wider. Kleine Grotten gruppieren sich um das zentrale Becken und schaffen ganz besondere Welten.

瓦尔斯温泉

瓦尔斯温泉酒店这一建筑近乎神圣，瑞士建筑师彼得·卒姆托以阿尔卑斯高山温泉的康复功能为重点，完成了一项富有创新性的设计。设计出色地展现了所处环境的特质——高山、原石、水和自然。在建设期间，共使用了60000块产自瓦尔斯的石英岩板材。它们层层堆叠而起，彼此间留有缝隙，不禁让人联想到洞穴。源自山中的温泉水，自然地融入了这一朴素的、人工的、近似岩石般的建筑中。在高墙之间回荡着风、脚步、滴水还有水流的声音。中心水池周边是小的“穴室”，它们自成天地，既美妙而又别致。

The outer pools, from which hot steam rises, seem almost surreal in the middle of the mountainous Swiss landscape.

Die Außenbecken, aus denen der warme Dampf steigt, wirken geradezu unwirklich inmitten der Schweizer Berglandschaft.

户外的温泉池，升腾的水蒸气，瑞士山间景色如同梦幻。

Smooth water surfaces, walls and lights are unified and bestow a truly mystical impression upon the room.

Die glatten Wasseroberflächen, Wände und Licht bilden eine Einheit und geben dem Raum einen geradezu mystischen Gesamteindruck.

光滑的水面、墙壁和光线浑然一体，让空间充满神秘感。

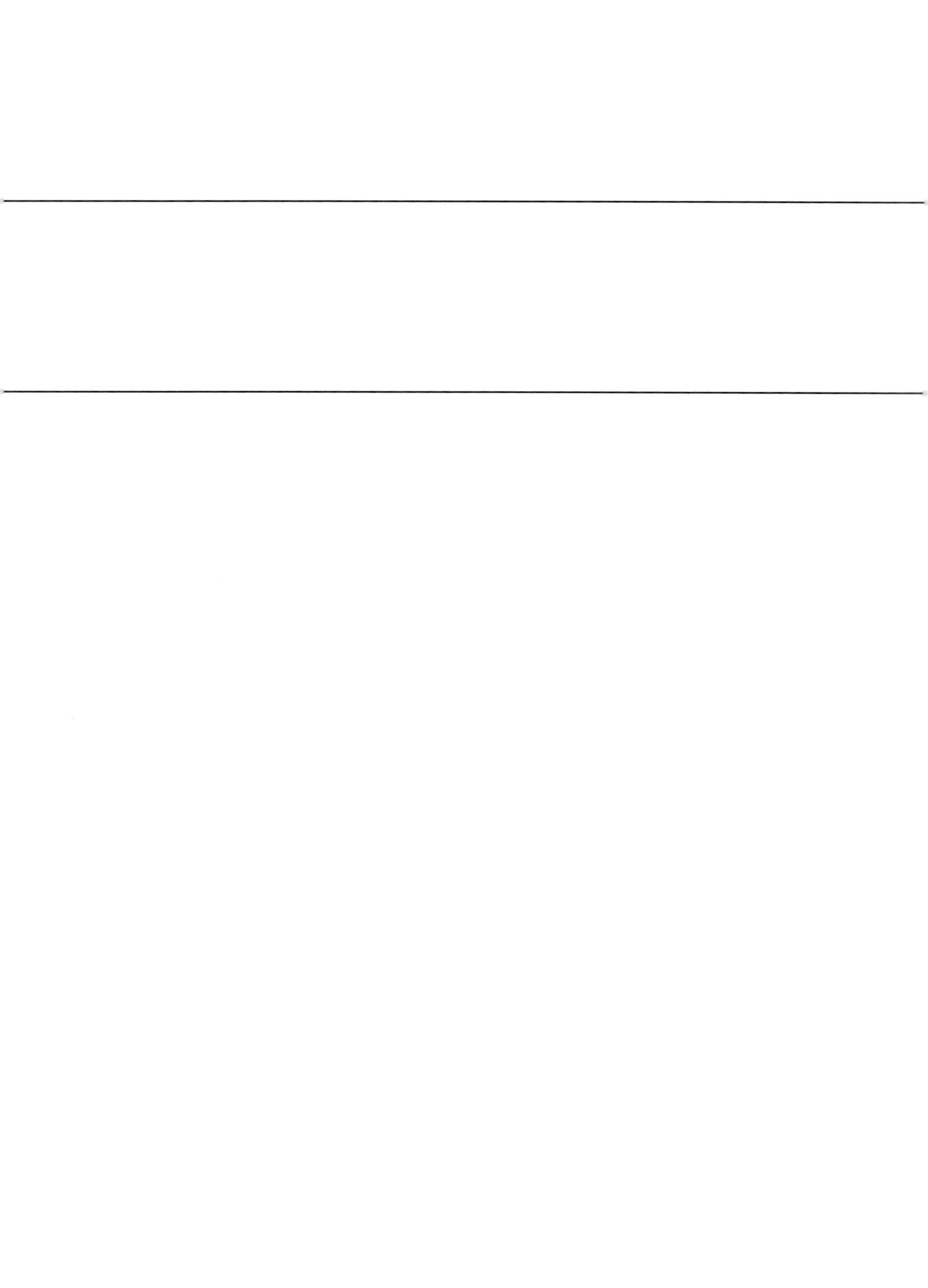

Directory | Verzeichnis

索引

3deluxe
www.3deluxe.de
Cyberhelvetia
摄影: *Emanuel Raab*

ACQ Architects
www.acq-architects.com
Dulwich Pool House
摄影: *Hufton+Crow/VIEW/artur*

Al Nakheel Properties
www.nakheel.com
The Palm Islands
摄影: *Al Nakheel Properties*

Tadao Ando
www.andotadao.com
Langen Foundation
摄影: *Langen Foundation*

Paul Andreu architecte
www.paul-andreu.com
National Grand Theater of China
摄影: *Josef Weiss/artur*

Atelier Dreiseitl
www.dreiseitl.de
Heiner-Metzger-Platz,
Queens Botanical Gardens
摄影: *Oleg Kuchar,*
Jeff Goldberg Esto

Behnisch Architekten
www.behnisch.com
Spa Baths Bad Aibling
摄影: *Behnisch Architekten*

Ernst Beneder
Haus H.
摄影: *Margherita Spiluttini*
Architekturfotografie

Mario Botta Architetto
www.botta.ch
San Carlino, SPA Bergoase
摄影: *Pino Mussi,*
Tschuggen Bergoase

Stephan Braunfels Architekten
www.braunfels-architekten.de
Paul-Löbe-Haus
摄影: *Thomas Raupach/artur*

BRT Bothe Richter
Teherani Architekten
www.brt.de
Dockland
摄影: *Joerg hempel/artur*

Santiago Calatrava
www.calatrava.com
City of Arts and Sciences
摄影: *Klaus Mellenthin,*
istockphoto

Coop Himmelb(l)au
www.coop-himmelblau.at
Art Museum Groningen
摄影: *archenova*

Delugan Meissl
Associated Architects
www.deluganmeissl.at
Haus Ray 1
摄影: *Hertha Hurnaus*
Photography

Denton Corker Marshall
www.dentoncorkermarshall.com
Webb Bridge
摄影: *Markus Bachmann,*
Erika Koch/artur

Diller + Scofidio
www.dillerscofidio.com
Blur Building
摄影: *Thomas Jantscher*
Architekturfotografie

Richard Hywel Evans
www.rhe.uk.com
Resort Zil Pasyon
摄影: *Per Aquum Resorts*

Architekten Förster Trabitzsch
www.architekten-ft.de
Floating Homes
摄影: *Klaus Frahm*

Gebr. Friedrich Schiffswerft
www.gfwerft.de
Living on Water
摄影: *Oliver Heissner*

GAD Architecture
www.gadarchitecture.com
Exploded House Project
摄影: *GAD Architecture*

Gehry Partners
www.foga.com
Guggenheim Museum
摄影: *Paul Raftery/VIEW/artur*

gmp Architekten von Gerkan,
Marg und Partner
www.gmp-architekten.de
Liquidrom
摄影: *Liquidrom*

Nicholas Grimshaw & Partners
www.ngrimshaw.co.uk
Bath Spa
摄影: *Edmund Sumner/VIEW/artur*

Gruppe Multipack
Aua extrema
摄影: *Thomas Jantscher*
Architekturfotografie

Gudmundur Jonsson Architects
www.internet.is/gudmundurjonsson
NORVEG Museum and Cultural Center
摄影: *Thomas Mayer Archive*

Zaha Hadid Architects
www.zaha-hadid.com
BMW Plant
摄影: *Werner Huthmacher/artur*

Atelier architecture
Kurt Hofmann
Hotel Palafitte
摄影: *Hotel Palafitte*

JSK Architekten
www.jsk.de
Medienhafen
摄影: *Martin Foddanu/artur*

Thomas Klumpp
Universum Science Center
摄影: *Universum Managementges. mbH, Karin Hessmann/artur*

Marcio Kogan
Architects+Designers
www.marciokogan.com.br
House Mirindiba
摄影: *Courtesy of Marcio Kogan*

Kengo Kuma & Associates
www.kkaa.co.jp
Baisouin Temple, Water/Glass Villa
摄影: *Edmund Sumner/VIEW/artur*

Rüdiger Lainer
www.lainer.at
Absberggasse School
摄影: *Margherita Spiluttini Architekturfotografie*

Richard Meier & Partners
www.richardmeier.com
Jesolo Lido Village
摄影: *Roland Halbe/artur*

MOS. office for architecture
www.mos-office.net
Floating House
摄影: *Florian Holzherr*

MVRDV
www.mvrdv.nl
Silodam
摄影: *MVRDV*

Neutelings Riedijk
www.neutelings-riedijk.com
The Sphinxes
摄影: *archenova*

Oscar Niemeyer
Publishing House Mondadori
摄影: *Roland Halbe/artur*

Tiago Oliveira
Hotel Estalagem da Ponta do Sol
摄影: *designhotels*

PURPUR.Architektur
& Vito Acconci
www.purpur.cc / www.acconci.com
Aiola Island Café
摄影: *Harry Schiffer, Aiola Island*

Rendel, Palmer and Tritton
Thames Flood Barrier
摄影: *istockphoto*

Miró Rivera Architects
www.mirorivera.com
Lake Austin Boat Dock
摄影: *Miró Rivera Architects*

RMP Stephan Lenzen
Landschaftsarchitekten
with Schmitz Architekten and Steidle+Partner
www.rmp-landschaftsarchitekten.de
T-Mobile Stadt
摄影: *C.R. Montag*

Sonu + Eva Shivdasani
Soneva Fushi
摄影: *designhotels*

studio lot
www.studiolot.de
Hofgut Hafnerleiten Guest Huts
摄影: *studio lot*

TANGRAM Architekten
www.tangramarchitecten.nl
Waterdwellings
摄影: *John Lewis Marshall*

tegnestuen vandkunsten ApS
www.vandkunsten.com
Torpedo Hall Apartments
摄影: *Jens M. Lindhe*

UNStudio
www.unstudio.com
Water Villas
摄影: *Markus Bachman*

Mies van der Rohe
Barcelona Pavilion
Thomas Spier/artur

Max Wehberg
www.buero-wehberg.de
Kraftwerk Spa Autostadt
摄影: *Jörn Hustedt*

WES & Partner
Landschaftsarchitekten
www.wesup.de
Penthouse Garden
摄影: *Jörn Hustedt*

Fridolin Welte
with Christian Lenz
St. Arbogast Water Pavilion
摄影: *Fridolin Welte*

Wilk-Salinas Architekten
www.gil-wilk.de
Badeschiff
摄影: *Thomas Spier/artur*

Peter Zumthor
Therme Vals
摄影: *Margherita Spiluttini Architekturfotografie*

著作权合同登记图字：01-2012-4629号

图书在版编目（CIP）数据

水与建筑设计（中英德文对照）／（德）菲舍尔著；周联译．—北京：中国建筑工业出版社，2014.7
（国外建筑设计案例精选）
ISBN 978-7-112-16826-2

Ⅰ.①水… Ⅱ.①菲…②周… Ⅲ.①理水（园林）－景观设计 Ⅳ.①TU986.4

中国版本图书馆CIP数据核字（2014）第095796号

Original title and ISBN:
Architecture Compact: Water-Wasser-Eau, 4884, ISBN 978-3-8331-5024-1

Editor: Joachim Fischer
Editorial coordination: Sabine Marinescu
Project coordination: Arne Alexander Klett
Layout: designdealer/büro für gestaltung
Imaging: designdealer/büro für gestaltung Produced by Klett Fischer
architecture + design publishing www.klett-fischer.com

责任编辑：孙立波 率 琦 白玉美 责任校对：陈晶晶 党 蕾

国外建筑设计案例精选
水与建筑设计
（中英德文对照）
［德］约阿希姆·菲舍尔 著
周 联 译
*
中国建筑工业出版社出版、发行（北京西郊百万庄）
各地新华书店、建筑书店经销
北京嘉泰利德公司制版
恒美印务（广州）有限公司印刷
*
开本：880×1230毫米 1/32 印张：9 字数：320千字
2014年11月第一版 2014年11月第一次印刷
定价：85.00元
ISBN 978-7-112-16826-2
（25608）